Approaches to Safe Nanotechnology

Managing the Health and Safety Concerns Associated with Engineered Nanomaterials

DEPARTMENT OF HEALTH AND HUMAN SERVICES
Centers for Disease Control and Prevention
National Institute for Occupational Safety and Health

Approaches to Safe Nanotechnology

Managing the Health and Safety Concerns Associated with Engineered Nanomaterials

DEPARTMENT OF HEALTH AND HUMAN SERVICES
Centers for Disease Control and Prevention
National Institute for Occupational Safety and Health

Disclaimer

Mention of any company or product does not constitute endorsement by the National Institute for Occupational Safety and Health (NIOSH). In addition, citations to Web sites external to NIOSH do not constitute NIOSH endorsement of the sponsoring organizations or their programs or products. Furthermore, NIOSH is not responsible for the content of these Web sites. All Web addresses referenced in this document were accessible as of the publication date.

Ordering Information

To receive documents or other information about occupational safety and health topics, contact NIOSH at

Telephone: **1–800–CDC–INFO** (1–800–232–4636)
TTY: 1–888–232–6348
E-mail: cdcinfo@cdc.gov

or visit the NIOSH Web site at **www.cdc.gov/niosh**.

For a monthly update on news at NIOSH, subscribe to *NIOSH eNews* by visiting **www.cdc.gov/niosh/eNews.**

DHHS (NIOSH) Publication No. 2009–125

March 2009

SAFER • HEALTHIER • PEOPLE™

Foreword

Nanotechnology—the manipulation of matter on a near-atomic scale to produce new structures, materials, and devices—offers the promise of unprecedented scientific advancement for many sectors, such as medicine, consumer products, energy, materials, and manufacturing. Nanotechnology has the power not only to improve existing technologies, but to dramatically enhance the effectiveness of new applications.

Research on the potential applications of nanotechnology continues to expand rapidly worldwide. New nanotechnology consumer products emerge at a rate of three to four per week. Over the course of the next decade, nanotechnology could have a $1 trillion impact on the global economy and employ two million workers—half of them residing in the U.S.

While nanomaterials present seemingly limitless possibilities, they bring with them new challenges to understanding, predicting, and managing potential safety and health risks to workers. The National Institute for Occupational Safety and Health (NIOSH) remains committed to protecting workers now and in the future, as nanotechnology applications and uses expand.

As part of these efforts, in October 2005, NIOSH released for public comment the draft document, Approaches to Safe Nanotechnology: An Information Exchange with NIOSH. Based on feedback received, NIOSH revised and updated the document in July 2006 and sought further public comment. This draft report has been widely cited, and the final version of the report should serve as a vital resource for stakeholders (including occupational safety and health professionals, researchers, policy makers, risk assessors, and workers in the industry) who wish to understand more about the safety and health implications of nanotechnology in the workplace.

With the publication of the Approaches to Safe Nanotechnology document, NIOSH hopes to: raise awareness of the occupational safety and health issues involved with nanotechnology; make recommendations on occupational safety and health best practices in the production and use of nanomaterials; facilitate dialogue between NIOSH and its external partners in industry, labor and academia; respond to requests for authoritative safety and health guidelines; and, identify information gaps and areas for future study and research.

As our knowledge of nanoscience increases, so too will our efforts to provide valuable guidance on the safe handling of nanoparticles and for protecting the lives and livelihoods of nanotechnology workers.

Christine M. Branche, Ph.D.
Acting Director, National Institute
  for Occupational Safety and Health
Centers for Disease Control and Prevention

Nanotechnology has the potential to dramatically improve the effectiveness of a number of existing consumer and industrial products and could have a substantial impact on the development of new products in all sectors, ranging from disease diagnosis and treatment to environmental remediation. Because of the broad range of possible nanotechnology applications, continued evaluation of the potential health risks associated with exposure to nanomaterials is essential to ensure their safe handling. Engineered nanoparticles are materials purposefully produced with at least one dimension between 1 and 100 nanometers. Nanoparticles* often exhibit unique physical and chemical properties that impart specific characteristics essential in making engineered materials, but little is known about what effect these properties may have on human health. Research has shown that the physicochemical characteristics of particles can influence their effects in biological systems. These characteristics include particle size, shape, surface area, charge, chemical properties, solubility, oxidant generation potential, and degree of agglomeration. Until the results from research studies can fully elucidate the characteristics of nanoparticles that may pose a health risk, precautionary measures are warranted.

NIOSH has developed this document to provide an overview of what is known about the potential hazards of engineered nanoparticles and measures that can be taken to minimize workplace exposures. Following is a summary of findings and key recommendations.

Potential Health Concerns

- The potential for nanomaterials to enter the body is among several factors that scientists examine in determining whether such materials may pose an occupational health hazard. Nanomaterials have the greatest potential to enter the body through the respiratory system if they are airborne and in the form of respirable-sized particles (nanoparticles). They may also come into contact with the skin or be ingested.

- Based on results from human and animal studies, airborne nanoparticles can be inhaled and deposit in the respiratory tract; and based on animal studies, nanoparticles can enter the blood stream, and translocate to other organs.

- Experimental studies in rats have shown that equivalent mass doses of insoluble incidental nanoparticles are more potent than large particles of similar composition in causing pulmonary inflammation and lung tumors. Results from in vitro cell culture studies with similar materials are generally supportive of the biological responses observed in animals.

- Experimental studies in animals, cell cultures, and cell-free systems have shown that changes in the chemical

*In an attempt at standardization of terminology, the International Organization for Standardization-Technical Committee 229 has used the term nanomaterial to describe engineered nanoparticles.

composition, crystal structure, and size of particles can influence their oxidant generation properties and cytotoxicity.

- Studies in workers exposed to aerosols of some manufactured or incidental microscopic (fine) and nanoscale (ultrafine) particles have reported adverse lung effects including lung function decrements and obstructive and fibrotic lung diseases. The implications of these studies to engineered nanoparticles, which may have different particle properties, are uncertain.

- Research is needed to determine the key physical and chemical characteristics of nanoparticles that determine their hazard potential.

Potential Safety Concerns

- Although insufficient information exists to predict the fire and explosion risk associated with powders of nanomaterials, nanoscale combustible material could present a higher risk than coarser material with a similar mass concentration given its increased particle surface area and potentially unique properties due to the nanoscale.

- Some nanomaterials may initiate catalytic reactions depending on their composition and structure that would not otherwise be anticipated based on their chemical composition.

Working with Engineered Nanomaterials

- Nanomaterial-enabled products such as nanocomposites, surface-coated materials, and materials comprised of nanostructures, such as integrated circuits,

are unlikely to pose a risk of exposure during their handling and use as materials of non-inhalable size. However, some of the processes used in their production (e.g., formulating and applying nanoscale coatings) may lead to exposure to nanomaterials, and the cutting or grinding of such products could release respirable-sized nanoparticles.

- Maintenance on production systems (including cleaning and disposal of materials from dust collection systems) is likely to result in exposure to nanoparticles if deposited nanomaterials are disturbed.

- The following workplace tasks can increase the risk of exposure to nanoparticles:

  — Working with nanomaterials in liquid media without adequate protection (e.g., gloves)

  — Working with nanomaterials in liquid during pouring or mixing operations, or where a high degree of agitation is involved

  — Generating nanoparticles in non-enclosed systems

  — Handling (e.g., weighing, blending, spraying) powders of nanomaterials

  — Maintenance on equipment and processes used to produce or fabricate nanomaterials and the cleaning-up of spills and waste material containing nanomaterials

  — Cleaning of dust collection systems used to capture nanoparticles

  — Machining, sanding, drilling, or other mechanical disruptions of materials containing nanoparticles

Exposure Assessment and Characterization

- Until more information becomes available on the mechanisms underlying nanomaterial toxicity, it is uncertain what measurement technique should be used to monitor exposures in the workplace. Current research indicates that mass and bulk chemistry may be less important than particle size and shape, surface area, and surface chemistry (or activity) for some nanostructured materials.

- Many of the sampling techniques that are available for measuring airborne nanoaerosols vary in complexity but can provide useful information for evaluating occupational exposures with respect to particle size, mass, surface area, number concentration, and composition. Unfortunately, relatively few of these techniques are readily applicable to routine exposure monitoring. NIOSH has initiated exposure assessment studies in workplaces that manufacture or use engineered nanoparticles (see Appendix *Nanoparticle Emission Assessment Technique for Identification of Sources and Releases of Engineered Nanomaterials*).

- Regardless of the metric or measurement method used for evaluating nanoaerosol exposures, it is critical that background nanoscale particle measurements be conducted before the production, processing, or handling of nanomaterials.

- When feasible, personal sampling is preferred to ensure an accurate representation of the worker's exposure, whereas area sampling (e.g., size-fractionated aerosol samples) and real-time (direct reading) exposure measurements may be more useful for evaluating the need for improvement of engineering controls and work practices.

Precautionary Measures

- Given the limited amount of information about health risks that may be associated with nanomaterials, taking measures to minimize worker exposures is prudent.

- For most processes and job tasks, the control of airborne exposure to nanoaerosols can be accomplished using a variety of engineering control techniques similar to those used in reducing exposure to general aerosols.

- The implementation of a risk management program in workplaces where exposure to nanomaterials exists can help to minimize the potential for exposure to nanoparticles. Elements of such a program should include the following:

  — Evaluating the hazard posed by the nanomaterial based on available physical and chemical property data, toxicology, or health-effects data

  — Assessing the worker's job task to determine the potential for exposure

  — Educating and training workers in the proper handling of nanomaterials (e.g., good work practices)

  — Establishing criteria and procedures for installing and evaluating engineering controls (e.g., exhaust ventilation) at locations where exposure to nanomaterials might occur

— Developing procedures for determining the need for and selecting proper personal protective equipment (e.g., clothing, gloves, respirators)

— Systematically evaluating exposures to ensure that control measures are working properly and that workers are being provided the appropriate personal protective equipment

• Engineering control techniques such as source enclosure (i.e., isolating the generation source from the worker) and local exhaust ventilation systems should be effective for capturing airborne nanoparticles. Current knowledge indicates that a well-designed exhaust ventilation system with a high-efficiency particulate air (HEPA) filter should effectively remove nanomaterials.

• The use of good work practices can help to minimize worker exposures to nanomaterials. Examples of good practices include cleaning of work areas using HEPA vacuum pickup and wet wiping methods, preventing the consumption of food or beverages in workplaces where nanomaterials are handled, providing hand-washing facilities, and providing facilities for showering and changing clothes.

• No guidelines are currently available on the selection of clothing or other apparel (e.g., gloves) for the prevention of dermal exposure to nanoaerosols. However, some clothing standards incorporate testing with nanometer-sized particles and therefore provide some indication of the effectiveness of protective clothing.

• Respirators may be necessary when engineering and administrative controls do not adequately prevent exposures. Currently, there are no specific limits for airborne exposures to engineered nanoparticles although occupational exposure limits exist for some larger particles of similar chemical composition. It should be recognized that exposure limits recommended for non-nanoscale particles may not be health protective for nanoparticle exposures (e.g., the OSHA Permissible Exposure Limit [PEL] for graphite may not be a safe exposure limit for carbon nanotubes). The decision to use respiratory protection should be based on professional judgment that takes into account toxicity information, exposure measurement data, and the frequency and likelihood of the worker's exposure. While research is continuing, preliminary evidence indicates that NIOSH-certified respirators will be useful for protecting workers from nanoparticle inhalation when properly selected and fit tested as part of a complete respiratory protection program.

Occupational Health Surveillance

Occupational health surveillance is an essential component of an effective occupational safety and health program. The unique physical and chemical properties of nanomaterials, the increasing growth of nanotechnology in the workplace, and information suggesting that exposure to some engineered nanomaterials can cause adverse health effects in laboratory animals all support consideration of an occupational health surveillance program for workers potentially exposed to engineered nanomaterials. Continued evaluation of toxicologic research and workers potentially exposed to engineered nanomaterials is needed to inform NIOSH and other groups regarding the appropriate components of occupational health surveillance for nanotechnology workers. NIOSH has formulated interim guidance relevant to medical screening (one component of an occupational health surveillance program) for nanotechnology workers (see NIOSH *Current Intelligence Bulletin Interim* *Guidance for Medical Screening and Hazard Surveillance for Workers Potentially Exposed to Engineered Nanoparticles* at www.cdc.gov/ niosh/review/public/115/). In this document NIOSH concluded that insufficient scientific and medical evidence now exist to recommend the specific medical screening of workers potentially exposed to engineered nanoparticles. However, NIOSH did recommend that hazard surveillance be conducted as the basis for implementing control measures.

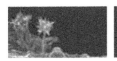

Contents

Acknowledgments

This report was developed by the scientists and staff of the National Institute for Occupational Safety and Health (NIOSH) who participate in the NIOSH Nanotechnology Research Center (NTRC). Paul Schulte is the manager and Charles Geraci, coordinator of the NIOSH NORA nanotechnology cross-sector program. Special thanks go to Ralph Zumwalde and Laura Hodson for writing and organizing the report and to Mark Methner for the development of the *Appendix: Nanoparticle Emission Assessment Technique for Identification of Sources and Releases of Engineered Nanomaterials.*

Others who contributed substantially to the writing and research described here include: Eileen Birch, Fred Blosser, Vincent Castranova, Brian Curwin, Douglas Evans, Pengfei Gao, Donna Heidel, Mark Hoover, John Howard, Vijia Karra, Bon-Ki Ku, Eileen Kuempel, Robert Mercer, Arthur Miller, Vladimir Murashov, Terri Pearce, Appavoo Rengasamy, Ronald Shaffer, Anna Shvedova, Petia Simeonova, Aleksandr Stefaniak, Douglas Trout, and Leonid Turkevich.

The NIOSH NTRC also acknowledges the contributions of Vanessa Becks and Gino Fazio for desktop publishing and graphic design, and Elizabeth Fryer for editing the report.

1 Introduction

Nanotechnology is the manipulation of matter on a near-atomic scale to produce new structures, materials, and devices. This technology has the ability to transform many industries and can be applied in many ways to areas ranging from medicine to manufacturing. Research in nanoscale technologies is growing rapidly worldwide. Lux Research [2007] projects that new emerging nanotechnology applications will affect nearly every type of manufactured product through the middle of the next decade, becoming incorporated into 15% of global manufacturing output, totaling $2.6 trillion in 2014.

Nanomaterials present new challenges to understanding, predicting, and managing potential health risks to workers. As with any material being developed, scientific data on the health effects in exposed workers are largely unavailable. **In the case of nanomaterials, the uncertainties are great because the characteristics of nanoparticles may be different from those of larger particles with the same chemical composition.** Safety and health practitioners recognize the critical lack of specific guidance on the safe handling of nanomaterials—especially now, when the degree of risk to exposed workers is unknown. In the meantime, the extensive scientific literature on airborne particles—including toxicology and epidemiological studies, measurement techniques, and engineering controls—provides the best available data from which to develop interim approaches for working safely with nanomaterials and to develop hypotheses for studies of new nanomaterials.

The National Institute for Occupational Safety and Health (NIOSH) is working in parallel with the development and implementation of commercial nanotechnology through (1) conducting strategic planning and research, (2) partnering with public- and private-sector colleagues from the United States and abroad, and (3), making information widely available. The NIOSH goal is to provide national and world leadership for incorporating research findings about the implications and applications of nanotechnology into good occupational safety and health practice for the benefit of all nanotechnology workers. NIOSH has developed a strategic plan for coordinating nanotechnology research and for use as a guide for enhancing the development of new research efforts (www.cdc.gov/niosh/topics/nanotech/strat_plan.html).

2 | Purpose

With the publication of this *Approaches to Safe Nanotechnology* document, NIOSH hopes to do the following:

- **Raise awareness** of the occupational safety and health issues being identified in the rapidly moving and changing science involving implications and applications of nanotechnology.

- **Use the best information available to make recommendations** on occupational safety and health practices in the production and use of nanomaterials (These recommendations will be updated as appropriate to reflect new information. They will address key components of occupational safety and health, including exposure monitoring, engineering controls, personal protective equipment, and administrative controls. They will draw from the ongoing NIOSH assessment of current best practices, technical knowledge, and professional judgment. Throughout the development of these guidelines, the utility of a hazard-based approach to risk assessment and control was evaluated and, where appropriate, recommendations are provided.)

- **Facilitate an exchange of information** between NIOSH and its external partners from ongoing research, including success stories, applications, and case studies.

- **Respond to requests** from industry, labor, academia, and other partners who are seeking science-based, authoritative guidelines.

- **Identify information gaps** where few or no data exist and where research is needed.

3 Scope

This document has been developed to provide a resource for stakeholders who wish to understand more about the safety and health implications and applications of nanotechnology in the workplace. The information and guidelines presented here are intended to aid in evaluating the potential hazard of exposure to engineered nanomaterials and to set the stage for the development of more comprehensive guidelines for reducing potential workplace exposures in the wide range of tasks and processes that use nanomaterials. The information in this document will be of specific interest to the following:

- Occupational safety and health professionals who must (1) understand how nanotechnology may affect occupational health and (2) devise strategies for working safely with nanomaterials

- Researchers working with or planning to work with engineered nanomaterials and studying the potential occupational safety and health impacts of nanomaterials

- Policy and decision makers in government agencies and industry

- Risk evaluation professionals

- People working with or potentially exposed to engineered nanomaterials in the workplace

Established safe work practices are generally based on an understanding of the hazards associated with the chemical and physical properties of a material. Engineered nanomaterials may exhibit unique properties that are related to their physical size, shape, structure, and chemical composition. Considerable uncertainty still exists as to whether these unique properties present occupational health risks. Current information about the potential adverse health effects of engineered nanomaterials, exposure assessment, and exposure control is limited. However, the large body of scientific literature that exists on exposures to and responses of animals and humans to ultrafine and other airborne particles may be useful in making preliminary assessments as to the health risks posed by engineered nanomaterials. **Until further information is available, interim safe working practices should be used based on the best available information.** The information and recommendations in this document are intended to aid in assessment of the potential hazard of engineered nanomaterials and to set the stage for the development of more comprehensive guidelines for reducing potential workplace exposures.

4 | Descriptions and Definitions

Nanotechnology involves the manipulation of matter at nanometer[†] scales to produce new materials, structures, and devices. The U.S. National Nanotechnology Initiative (see http://nano.gov/html/facts/whatIsNano.html) defines a technology as nanotechnology only if it involves all of the following:

- Research and technology development involving structures with at least one dimension in the range of 1–100 nanometers (nm), frequently with atomic/molecular precision

- Creating and using structures, devices, and systems that have unique properties and functions because of their nanoscale dimensions

- The ability to control or manipulate on the atomic scale

Nanotechnology is an enabling technology that offers the potential for unprecedented advances in many diverse fields. The ability to manipulate matter at the atomic or molecular scale makes it possible to form new materials, structures, and devices that exploit the unique physical and chemical properties associated with nanoscale structures. The promise of nanotechnology goes far beyond extending the use of current materials. New materials and devices with intricate and closely engineered structures will allow for (1) new directions in optics, electronics, and optoelectronics, (2) development of new medical imaging and treatment technologies, and (3) production of advanced materials with unique properties and high-efficiency energy storage and generation.

Although nanotechnology-based products are generally thought to be at the precompetitive stage, an increasing number of products and materials are becoming commercially available. These include nanoscale powders, solutions, and suspensions of nanoscale materials as well as composite materials and devices having a nanostructure. Nanoscale products and materials are increasingly used in optoelectronic, electronic, magnetic, medical imaging, drug delivery, cosmetic, catalytic, and materials applications. New nanotechnology consumer products are coming on the market at the rate of three to four per week, a finding based on the latest update to the nanotechnology consumer product inventory maintained by the Project on Emerging Nanotechnologies (PEN)[‡] (www.nanotechproject.org/inventories/consumer). The number of consumer products using nanotechnology has grown from 212 to 609 since PEN launched the world's first online inventory of manufacturer-identified nanotech goods in March 2006.

According to Lux Research [2007], in 2006, governments, corporations, and venture capitalists worldwide spent $11.8 billion on nanotechnology research and development (R&D), which was up 13% from 2005. By 2014, Lux estimates $2.6 trillion in manufactured goods

---

[†]1 nanometer (nm) = 1 billionth of a meter (10^{-9}).

[‡]The Project on Emerging Nanotechnologies was established in April 2005 as a partnership between the Woodrow Wilson International Center for Scholars and the Pew Charitable Trusts.

Figure 4–1. Photomicrographs of airborne exposure to ultrafine (nanoscale) particles of welding fumes, diesel exhaust, and cerium oxide

will incorporate nanotechnology—or about 15% of total global output.

4.1 Nano-objects

The International Organization for Standardization Technical Committee 229 (Nanotechnologies) is developing globally recognized nomenclature and terminology for nanomaterials. According to ISO/TS 27687:2008, *nano-object* is defined as material with one, two, or three external dimensions in the size range from approximately 1–100 nm. Subcategories of nano-object are (*1*) *nanoplate*, a nano-object with one external dimension at the nanoscale; (*2*) *nanofiber*, a nano-object with two external dimensions at the nanoscale with a nanotube defined as a hollow nanofiber and a nanorod as a solid nanofiber; and (*3*) *nanoparticle*, a nano-object with all three external dimensions at the nanoscale. Nano-objects are commonly incorporated in a larger matrix or substrate referred to as a *nanomaterial*. Nano-objects may be suspended in a gas (as a nanoaerosol), suspended in a liquid (as a colloid or nanohydrosol), or embedded in a matrix (as a nanocomposite).

The precise definition of particle diameter depends on particle shape as well as how the diameter is measured. Particle morphologies may vary widely at the nanoscale. For instance, carbon fullerenes represent nano-objects with identical dimensions in all directions (i.e., spherical), whereas single-walled carbon nanotubes (SWCNTs) typically form convoluted, fiber-like nano-objects. Many regular but nonspherical particle morphologies can be engineered at the nanoscale, including flower- and belt-like structures. Please see www.nanoscience.gatech.edu/zlwang/research.html for examples of some nanoscale structures.

4.2 Ultrafine Particles

The term *ultrafine particle* has traditionally been used by the aerosol research and occupational and environmental health communities to describe airborne particles smaller than 100 nm in diameter. *Ultrafine* is frequently used in the context of nanometer-diameter particles that have not been intentionally produced but are the incidental products of processes involving combustion, welding, or diesel engines (see Figure 4–1). The term *nanoparticle* is frequently used with respect to particles demonstrating size-dependent physicochemical properties, particularly from a materials science perspective. **The two terms are sometimes used to differentiate between engineered** (*nanoparticle*) and incidental (*ultrafine*) **nanoscale particles**.

It is currently unclear whether the use of source-based definitions of nanoparticles and ultrafine particles is justified from a safety and health perspective. This is particularly the case where data on non-engineered, nanometer-diameter particles are of direct relevance to the impact of engineered particles. An attempt has been made in this document to follow the general convention of preferentially using *nanoparticle* in the context of intentionally produced or engineered particles and *ultrafine* in the context of incidentally produced particles (e.g., combustion products). However, this does not necessarily imply specific differences in the properties of these particles as related to hazard assessment, measurement, or control of exposures, and this remains an active area of research. *Nanoparticle* and *ultrafine particle* are not rigid definitions. For example, since the term *ultrafine* has been in existence longer, some intentionally produced particles with primary particle sizes in the nanosize range (e.g., TiO_2) are often called *ultrafine* in the literature.

4.3 Engineered Nanoparticles

Engineered nanoparticles are intentionally produced, whereas *ultrafine particles* (often referred to as *incidental nanoparticles*) are typically byproducts of processes such as combustion and vaporization. Engineered nanoparticles are designed with very specific properties or compositions (e.g., shape, size, surface properties, and chemistry). Incidental nanoparticles are generated in a relatively uncontrolled manner and are usually physically and chemically heterogeneous compared with engineered nanoparticles.

4.4 Nanoaerosol

A *nanoaerosol* is a collection of nanoparticles suspended in a gas. The particles may be present as discrete nano-objects, or as aggregates or agglomerates of nano-objects. These agglomerates may have diameters larger than 100 nm. In the case of an aerosol consisting of micrometer-diameter particles formed as agglomerates of nano-objects, the definition of nanoaerosol is open to interpretation. It is generally accepted that if the nanostructure associated with the nano-object is accessible (through physical, chemical, or biological interactions), then the aerosol may be considered a nanoaerosol. However, if the nanostructure within individual micrometer-diameter particles does not directly influence particle behavior (for instance, if the nanoparticles were inaccessibly embedded in a solid matrix), the aerosol would not be described as a nanoaerosol.

4.5 Agglomerate

An *agglomerate* is a group of nanoparticles held together by relatively weak forces, including van der Waals forces, electrostatic forces, and surface tension [ISO 2006].

4.6 Aggregate

An *aggregate* is a heterogeneous particle in which the various components are held together by relatively strong forces, and thus not easily broken apart [ISO 2006]. Aggregated nanoparticles would be an example of a nanostructured material.

Nanotechnology is an emerging field. As such, there are many uncertainties as to whether the unique properties of engineered nanomaterials (which underpin their commercial and scientific potential) also pose occupational health risks. These uncertainties arise because of gaps in knowledge about the factors that are essential for predicting health risks—factors such as routes of exposure, translocation of materials once they enter the body, and interaction of the materials with the body's biological systems. The potential health risk following exposure to a substance is generally associated with the magnitude and duration of the exposure, the persistence of the material in the body, the inherent toxicity of the material, and the susceptibility or health status of the person exposed. More data are needed on the health risks associated with exposure to engineered nanomaterials. Results of existing studies in animals and humans on exposure and response to ultrafine or other respirable particles provide a basis for preliminary estimates of the possible adverse health effects from exposures to similar engineered materials on a nanoscale. Experimental studies in rodents and cell cultures have shown that the toxicity of ultrafine or nanoparticles is greater than that of the same mass of larger particles of similar chemical composition [Oberdörster et al. 1992, 1994a, b; Lison et al. 1997; Tran et al. 1999, 2000; Brown et al. 2001; Barlow et al. 2005; Duffin et al. 2007]. In addition to particle surface area, other particle characteristics may influence toxicity, including surface functionalization or coatings, solubility, shape, and the ability to generate oxidant species and to adsorb biological proteins or bind to receptors [Duffin et al. 2002; Oberdörster et al. 2005a; Maynard and Kuempel 2005; Donaldson et al. 2006]. More research is needed on the influence of particle properties on interactions with biological systems and the potential for adverse effects. International research strategies for evaluating the safety of nanomaterials are actively being developed through cooperative efforts [Thomas et al. 2006].

Existing toxicity information about a given material of larger particle size can provide a baseline for anticipating the possible adverse health effects that may occur from exposure to a nanoscale material that has some of the same physicochemical properties (e.g., chemistry, density). However, predicting the toxicity of an engineered nanomaterial based on its physicochemical properties may not provide an adequate level of protection.

5.1 Exposure Routes

Inhalation is the most common route of exposure to airborne particles in the workplace. The deposition of discrete nano-objects in the respiratory tract is determined by the particle's aerodynamic or thermodynamic diameter (i.e., the particle shape and size). Agglomerates of nano-objects will deposit according to the diameter of the agglomerate, not constituent nano-objects. Research is ongoing to determine the physical factors that contribute to the agglomeration and de-agglomeration of nano-objects in air, suspended in aqueous media, or once in contact with lung lining fluid and/or biological proteins. Evidence indicates that the degree of agglomeration can affect the toxicity of inhaled nano-objects [Shvedova et al. 2007].

Discrete nanoparticles are deposited in the lungs to a greater extent than larger respirable particles [ICRP 1994], and deposition increases with exercise due to increase in breathing rate and change from nasal to mouth breathing [Jaques and Kim 2000; Daigle et al. 2003] and among persons with existing lung diseases or conditions (e.g., asthma, emphysema) [Brown et al. 2002]. Based on animal studies, discrete nanoparticles may enter the bloodstream from the lungs and translocate to other organs [Takenaka et al. 2001; Nemmar et al. 2002; Oberdörster et al. 2002].

Discrete nanoparticles (35–37-nm median diameter) that deposit in the nasal region may be able to enter the brain by translocation along the olfactory nerve, as was observed in rats [Oberdörster et al. 2004; Oberdörster et al. 2005a; Elder et al. 2006]. The transport of insoluble particles from 20–500 nm-diameter to the brain via sensory nerves (including olfactory and trigeminus) was reported in earlier studies in several animal models [De Lorenzo 1970; Adams and Bray 1983; Hunter and Dey 1998]. This exposure route for nanoparticles and to nanoscale biological agents has not been studied in humans.

Some studies suggest that nanomaterials could potentially enter the body through the skin during occupational exposure. Tinkle et al. [2003] have shown that particles smaller than 1 μm in diameter may penetrate into mechanically flexed skin samples. A more recent study reported that nanoparticles with varying physicochemical properties were able to penetrate the intact skin of pigs [Ryman-Rasmussen et al. 2006]. These nanoparticles were quantum dots of different size, shape, and surface coatings. They were reported to penetrate the stratum corenum barrier by passive diffusion and localize within the epidermal and dermal layers within 8–24 hours. The dosing solutions were 2- to 4-fold dilutions of quantum dots as commercially supplied and thus represent occupationally relevant doses.

At this time, it is not fully known whether skin penetration of nanoparticles would result in adverse effects in animal models. However, topical application of raw SWCNT to nude mice has been shown to cause dermal irritation [Murray et al. 2007]. Studies conducted in vitro using primary or cultured human skin cells have shown that both SWCNT and multi-walled carbon nanotubes (MWCNT) can enter cells and cause release of pro-inflammatory cytokines, oxidative stress, and decreased viability [Monteiro-Riviere et al. 2005; Shvedova et al. 2003]. It remains unclear, however, how these findings may be extrapolated to a potential occupational risk, given that additional data are not yet available for comparing the cell model studies with actual conditions of occupational exposure. Research on the dermal exposure of nanomaterials is ongoing (www.uni-leipzig.de/~nanoderm/).

Ingestion can occur from unintentional hand to mouth transfer of materials; this has been found to happen with traditional materials, and it is scientifically reasonable to assume that it also could happen during handling of nanomaterials. Ingestion may also accompany inhalation exposure because particles that are cleared from the respiratory tract via the mucociliary escalator may be swallowed [ICRP 1994]. Little is known about possible adverse effects from the ingestion of nanomaterials.

5.2 Effects Seen in Animal Studies

Experimental studies in rats have shown that at equivalent mass doses, insoluble ultrafine particles are more potent than larger particles of similar composition in causing pulmonary inflammation, tissue damage, and lung tumors [Lee et al. 1985; Oberdörster and Yu 1990; Oberdörster et al. 1992, 1994a,b; Heinrich et al. 1995; Driscoll 1996; Lison et al. 1997; Tran et al. 1999, 2000; Brown et al. 2001; Duffin et al. 2002; Renwick et al. 2004; Barlow et al. 2005]. These studies have shown that for poorly-soluble low toxicity (PSLT) particles, the dose-response relationships are consistent across particle sizes when dose is expressed as particle surface area. In addition to particle size and surface area, studies have shown that other particle characteristics can influence toxicity. For example, although the relationship between particle surface area dose and pulmonary inflammation is consistent among PSLT particles, crystalline silica is much more inflammogenic than PSLT particles at a given surface area dose [Duffin et al. 2007].

Reactive oxidant generation on the particle surface is an important factor influencing lung response to particles, which can be related to crystal structure. A recent study of the lung effects of rats dosed with either ultrafine *anatase* titanium dioxide (TiO_2) or ultrafine *rutile* TiO_2 showed that the *anatase* TiO_2 had more reactive surfaces and caused greater pulmonary inflammation and cell proliferation in the lungs of rats [Warheit et al. 2007]. In a cell-free assay designed to investigate the role of surface area and crystal structure on particle reactive oxygen species (ROS)-generation, Jiang et al. [2008] observed that size, surface area, and crystal structure all contribute to ROS generation.

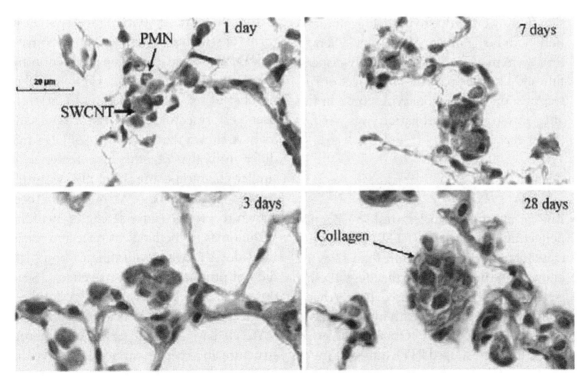

Figure 5–1. Formation of collagen following deposition of SWCNTs in the lungs of mice

Oxidant generation was apparently associated with the number of defective sites per surface area, which varied in nanoparticles in some size ranges [Jiang et al. 2008].

These studies indicate that for nanoparticles with similar properties (e.g., PSLT), the toxicity of a given mass dose will increase with decreasing particle size due to the increasing surface area. However, the dose-response relationship may differ for particles with different chemical composition and other properties. Consistent with these findings, a recent pulmonary instillation study with rats dosed with either fine or ultrafine TiO_2 reported no significant difference in lung responses when compared to controls, while crystalline silica caused more severe lung responses at the same dose [Warheit et al. 2006]. However, Warheit et al. [2006] were unable to adequately test the hypotheses about the relationship between particle surface area dose and toxicity because the diameters of the fine and ultrafine TiO_2-instilled particles did not significantly differ due to particle agglomeration, both being in excess of 2 μm. When efforts were made to more effectively disperse fine and ultrafine particles, the effect of surface area on the pulmonary response in rats after intratracheal instillation was verified [Sager et al. 2008].

5.2.1 Polytetrafluoroethylene fume

Among ultrafine particles, freshly generated polytetrafluoroethylene (PTFE) fume (generated at temperatures of more than 425°C) is known to be highly toxic to the lungs. Freshly generated PTFE fume caused hemorrhagic pulmonary edema and death in rats exposed to less than 60 μg/m^3 [Oberdörster et al. 1995]. In contrast, aged PTFE fume was much less toxic and did not result in mortality. This low toxicity was attributed to the increase

in particle size from accumulation and to changes in surface chemistry [Johnston et al. 2000; Oberdörster et al. 2005a]. Human case studies have reported pulmonary edema in workers exposed to PTFE fume and an accidental death in a worker when an equipment malfunction caused overheating of the PTFE resin and release of the PTFE pyrolysis products in the workplace [Goldstein et al. 1987; Lee et al. 1997]. While PTFE fume differs from engineered nanoparticles, these studies illustrate properties of ultrafine particles that have been associated with an acute toxic hazard. Enclosed processes and other engineering controls appear to have been effective at eliminating worker exposures to PTFE fume in normal operations, and thus may provide examples of control systems that may be implemented to prevent exposure to nanoparticles that may have similar properties.

5.2.2 Carbon nanotubes

Carbon nanotubes (CNT) are specialized forms or structures of engineered nanomaterials that have had increasing production and use [Donaldson et al. 2006]. Consequently, a number of toxicologic studies of CNT have been performed in recent years. These studies have shown that the toxicity of CNT may differ from that of other nanomaterials of similar chemical composition. For example, single-walled CNTs (SWCNT) have been shown to produce adverse effects including granulomas in the lungs of mice and rats at mass doses at which ultrafine carbon black did not produce these adverse effects [Shvedova et al. 2005; Lam et al. 2004]. While both SWCNTs and carbon black are carbon-based, SWCNTs have a unique, convoluted, fibrous structure and specific surface chemistry that offers excellent electrical conductive properties. How these characteristics may influence

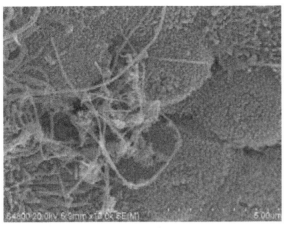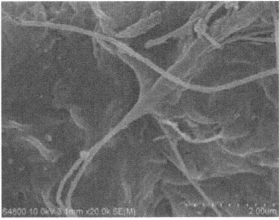

Figure 5–2. Deposition and clearance of MWCNTs from the conducting airways of mice following inhalation exposure

toxicity is not known. Carbon nanotubes may contain metal catalysts as byproducts of their production, which could contribute to their toxicity, or the CNTs may provide a structure that promotes fibroblast cell growth [Wang et al. 2008].

In a study of SWCNTs instilled into the lungs of rats, multi-focal granulomas (without transient inflammation or persistent lesions) were observed at doses of 1 or 5 mg/kg body weight [Warheit et al. 2004]. In a study of mice instilled with one of several types of SWCNTs (i.e., raw, purified, iron-containing, and nickel-containing) at doses of 0.1 or 0.5 mg/mouse (approximately 3 or 16 mg/kg body weight), dose-dependent epithelioid granulomas were observed at 7 days, which persisted at 90 days [Lam et al. 2004, 2006]. Both the raw and purified forms produced interstitial inflammation, while mortality (5/9 mice) was observed in the high dose group of the Ni-containing SWCNT.

NIOSH researchers recently reported adverse lung effects following pharyngeal aspiration of SWCNTs in mice using doses between 10–40 μg/mouse (approximately 0.5–2 mg/kg body weight) [Shvedova et al. 2005]. The findings showed that exposure to SWCNTs in mice lead to transient pulmonary inflammation, oxidative stress, decrease in pulmonary function, decrease in bacterial clearance, and early onset of interstitial fibrosis. Deposition of agglomerates resulted in development of granulomas, while deposition of dispersed nanotube structures in the aspirated suspension resulted in the rapid development of interstitial fibrosis (within 7 days), which progressed over a 30–60 day post-exposure period [Shvedova et al. 2005; Mercer et al. 2008]. Evidence indicates that when efforts were made to more fully disperse the SWCNT and obtain smaller structures in the aspiration suspension, fewer granulomas occurred but a 4-fold more potent interstitial fibrotic response was observed [Mercer et al. 2008].

Exposure to SWCNT has been observed to be more fibrogenic than an equal mass of either ultrafine carbon black or fine quartz [Shvedova et al. 2005; Lam et al. 2004]. Based on their findings in mice, Shvedova et al. [2005] estimated that workers may be at risk of developing lung lesions if they were exposed to SWCNT over a period of 20 days at the current

OSHA PEL for graphite (5 mg/m³). Lam et al. [2004, 2006] provided similar estimates and suggested that the graphite PEL should not be used (e.g., on MSDS) as a safe concentration for workers exposed to CNTs. Compared to instillation, the pharyngeal aspiration technique may approximate more closely the particle deposition that occurs during inhalation. Inhalation studies of CNTs may provide more definitive information about their potential toxicity in humans [Donaldson et al. 2006]. Recently, NIOSH scientists designed a system to generate an aerosol of SWCNT for a rat inhalation study [Baron et al. 2008]. Results of the inhalation exposure to SWCNT [Shvedova et al. 2008] were qualitatively similar to those of the aspiration study [Shvedova et al. 2004] with a 4-fold more potent interstitial fibrotic response similar to that reported by Mercer et al. [2008]. Another NIOSH study found markers of inflammation in the lung, aorta, and heart tissues of ApoE-/- mice after a single intra-pharyngeal instillation dose of SWCNT (10 and 40 μg/mouse) and accelerated plaque formation after repeated doses (20 μg/mouse once every other week for 8 weeks in mice fed an atherogenic diet) [Li et al. 2007].

MWCNTs were recently studied by intratracheal instillation in Sprague-Dawley rats receiving 0.5, 2, or 5 mg (approximately 2, 9, or 22 mg/kg body weight) of either ground MWCNT or unground MWCNT [Muller et al. 2005]. Both forms produced pulmonary inflammation and fibrosis. Rats that received ground MWCNT showed greater dispersion in the lungs, and fibrotic lesions were observed in the deep lungs (alveolar region). In rats treated with MWCNT (not ground) fibrosis showed mainly in their airways rather than in their lungs. The biopersistence of the unground MWCNT was greater than that of the ground MWCNT, with 81% vs. 36%, respectively, remaining in the lungs at day 60.

At an equal mass dose, ground MWCNT produced a similar inflammatory and fibrogenic response as chrysotile asbestos and a greater response than ultrafine carbon black [Muller et al. 2005]. Effects from the vehicle (1% Tween 80) used for administering ground and unground MWCNT to rats were not reported; the control group used in the study was exposed to only saline. NIOSH scientists have exposed mice by aspiration to MWCNT suspended in a simulated alveolar lining fluid rather than Tween 80. Control studies show that this suspension medium was not inflammatory and did not mask the biological activity of the particle surface. Data indicate that aspiration of dispersed MWCNT produced pulmonary inflammation, which peaked 7 days post exposure. The inflammatory response to MWCNT was greater than the inflammatory response to SWCNT [Sriram et al. 2007].

Two recent studies investigated the hypotheses that CNTs can behave like asbestos. In the first study, Takagi et al. [2008] administered to p53 (+/-) mice MWCNT, fullerene, or crocidolite asbestos by intraperitoneal injection at doses of 3 mg/mouse. The average width of the MWCNT was approximately 100 nm, and approximately 28% of the particles were longer than 5 μm. The particle number concentrations of MWCNT and crocidolite were 1×10^9 and 1×10^{10} (in 1-ml suspensions), respectively, although the MWCNT sample was also reported to contain mainly large aggregates, indicating that the number of MWCNT fibers was vastly underestimated and much larger than for the asbestos exposure. At the termination of the study (25 weeks), mesothelial responses in the MWCNT-treated mice included moderate to severe fibrous peritoneal adhesion and peritoneal tumors. The asbestos-treated mice had similar responses but to a lesser extent, while the fullerene-treated group did not show these responses. Mesothelioma was considered by the authors as the primary cause of death, and constriction of the ileus due to severe peritoneal adhesion was considered to be the second major cause of death, suggesting that 3 mg/mouse exceeded the maximum tolerated dose of MWCNT. Whether mesotheliomia was a primary cause of death is somewhat speculative.

In a second study, Poland et al. [2008] administered to mice either MWCNT (two short and two long CNT samples), nanoscale carbon black, or amosite (short or long) at doses of 50 μg/mouse by intraperitoneal injection. The short CNTs were 10 nm or 15 nm in width, with no fibers larger than 15 μm in length detected; the long CNTs were 85 nm or 165 nm in width, and 24% or 84%, respectively, were larger than 15 μm in length (the percentage of fibers longer than 5 μm was not reported). After either 24 hours or 7 days, the long MWCNT caused inflammation and granulomatous lesions that were qualitatively and quantitatively similar to that caused by the long asbestos. The short, low-aspect-ratio, tangled aggregates of MWCNT did not produce these responses at the doses used in this study. Additional studies are needed to determine if this inflammatory response to MWCNT would be persistent and result in tumors of the abdominal wall. Additionally, the potential for migration of MWCNT through the lungs to the mesothelium after inhalation requires investigation. Long-term studies are also needed to determine whether CNTs can cause cancer such as mesothelioma in laboratory animals, including exposures by typical routes in humans (i.e., inhalation, dermal penetration, and ingestion) and at doses that include those equivalent to potential workplace exposures.

These studies indicate the need for more data on exposures of workers to CNTs. Maynard et al. [2004] reported relatively low short-term (approximately 30 min) airborne mass concentrations of SWCNT (0.007–0.053 mg/m^3) in a laboratory production facility. A recent study by Han et al. [2008] reported total airborne mass concentrations of MWCNT from 0.21–0.43 mg/m^3 (4–6-hr sampling) in a laboratory research facility prior to use of engineering control measures; after implementing controls, the concentration decreased to nondetectable. Workers could also be exposed to ground CNTs used in polymer composites and other matrices or during cutting, grinding, or polishing of these materials. Given that exposure to SWCNT and MWCNT causes interstitial fibrosis and pulmonary inflammation, respectively, in rodent lungs at relatively low mass doses, it is prudent to minimize worker exposure to airborne CNTs (see Chapter 8 *Guidelines for Working with Engineered Nanomaterials*).

5.3 Observations from Epidemiological Studies Involving Fine and Ultrafine Particles

Epidemiological studies in workers exposed to aerosols including fine and ultrafine particles have reported lung function decrements, adverse respiratory symptoms, chronic obstructive pulmonary disease, and fibrosis [Kreiss et al. 1997; Gardiner et al. 2001; Antonini 2003]. In addition, some studies have found lung disease including elevated lung cancer and neurological effects among workers exposed to certain ultrafine particles (i.e., diesel exhaust particulate) [Steenland et al. 1998; Garshick et al. 2004, 2006; Hart et al. 2006] or welding fumes [Antonini 2003; Park et al. 2006; Ambroise et al. 2007; Bowler et al. 2007]. The implications of these studies to engineered nanomaterials, which may have different particle properties, are uncertain. Studies of airborne particles and fibers in the workplace do provide relevant background information about the particle-related lung diseases and mechanisms, and some limited quantitative estimates of exposures and risk of adverse health effects. As such, these studies provide a point of reference, including baseline information and estimates regarding possible health risks of exposure to other nanoscale particles depending on the extent to which the exposure conditions and particle-biological interactions may be similar.

Epidemiological studies in the general population have also shown associations between particulate air pollution and increased morbidity and mortality from respiratory and cardiovascular diseases [Dockery et al. 1993; HEI 2000; Pope et al. 2002, 2004]. Some epidemiological studies have shown adverse health effects associated with exposure to the ultrafine particulate fraction of air pollution [Peters et al. 1997, 2004; Penttinen et al. 2001; Ibald-Mulli et al. 2002; Timonen et al. 2004; Ruckerl et al. 2006] although uncertainty exists about the role of ultrafine particles relative to other air pollutants in causing the observed adverse health effects. The associations in these studies have been based on measurements of the particle number or mass concentrations of particles within certain size fractions (e.g., particulate matter with diameter of 2.5 μm and smaller [$PM_{2.5}$]). In an experimental study of healthy and asthmatic subjects inhaling ultrafine carbon particles, changes were observed in the expression of adhesion molecules by blood leukocyte, which may relate to possible cardiovascular effects of ultrafine particle exposure [Frampton et al. 2006]. Short-

term diesel exhaust exposure (0.3 mg/m^3 for 1 hr) in healthy volunteers was associated with mild systemic inflammation and impaired endothelial-dependent vasodilation [Törnqvist et al. 2007].

5.4 Hypotheses from Animal and Epidemiological Studies

The existing literature on particles and fibers provides a scientific basis from which to evaluate the potential hazards of engineered nanomaterials. While the properties of engineered nanomaterials can vary widely, the basic physicochemical and toxicokinetic principles learned from the existing studies are relevant to understanding the potential toxicity of nanomaterials. For example, it is known from studies in humans that a greater proportion of inhaled nanoparticles will deposit in the respiratory tract (both at rest and with exercise) compared to larger particles [ICRP 1994; Jaques and Kim 2000; Daigle et al. 2003; Kim and Jaques 2004]. It is also known from studies in animals that nanoparticles in the lungs can be translocated to other organs in the body; how the chemical and physical properties of the nanoparticles influence this translocation is not completely known [Takenaka et al. 2001; Kreyling et al. 2002; Oberdörster et al. 2002, 2004; Semmler et al. 2004; Geiser et al. 2005]. Due to their small size, nanoparticles can cross cell membranes and interact with subcellular structures such as mitochondria, where they have been shown to cause oxidative damage and to impair function of cells in culture [Möller et al. 2002, 2005; Li et al. 2003; Geiser et al. 2005]. Nanoparticles have also been observed inside cell nuclei [Porter et al. 2007a, b]. Animal studies have shown that nanoparticles are more biologically active due to their greater surface area per mass compared with larger-sized particles of the same chemistry [Oberdörster et al. 1992; 1994a,b; 2005a; Driscoll 1996; Lison et al. 1997; Brown et al. 2001; Duffin et al. 2002; Renwick et al. 2004; Barlow et al. 2005; Sager et al. 2008]. While this increased biological activity is a fundamental component to the utility of nanoparticles for industrial, commercial, and medical applications, the consequences of unintentional exposures of workers to nanoparticles are uncertain.

Research reported from laboratory animal studies and from epidemiological studies have lead to hypotheses regarding the potential adverse health effects of engineered nanomaterials. These hypotheses are based on the scientific literature of particle exposures in animals and humans. This literature has been recently reviewed [Donaldson et al. 2005; Maynard and Kuempel 2005; Oberdörster et al. 2005a, Donaldson et al. 2006; Kreyling et al. 2006]. In general, the particles used in past studies have not been characterized to the extent recommended for new studies in order to more fully understand the physicochemical properties of the particles that influence toxicity [Oberdörster et al. 2005b; Thomas et al. 2006]. As this research continues, more data will become available to support or refute the following hypotheses for engineered nanoparticles.

Hypothesis 1: Exposure to engineered nanoparticles is likely to cause adverse health effects similar to ultrafine particles that have similar physical and chemical characteristics.

Studies in rodents and humans support the hypothesis that exposure to ultrafine particles poses a greater respiratory hazard than exposure to the same mass of larger particles with a similar chemical composition. Studies of existing particles have shown adverse health effects in workers exposed to ultrafine particles (e.g., diesel exhaust particulate, welding fumes), and animal studies have shown that ultrafine particles are more inflammogenic and tumorigenic in the lungs of rats than an equal mass of larger particles of similar composition [Oberdörster and Yu 1990; Driscoll 1996; Tran et al. 1999, 2000]. **If engineered nanoparticles have the same physicochemical characteristics that are associated with reported effects from ultrafine particles, they may pose the same health concerns.**

Although the physicochemical characteristics of ultrafine particles and engineered nanoparticles can differ, the toxicologic and dosimetric principles derived from available studies may be relevant to postulating the health concerns for newly engineered particles. The biological mechanisms of particle-related lung diseases (i.e., oxidative stress, inflammation, and production of cytokines, chemokines, and cell growth factors) [Mossman and Churg 1998; Castranova 2000; Donaldson and Tran 2002] appear to be a consistent lung response for respirable particles including ultrafine or engineered nanoparticles [Donaldson et al. 1998; Donaldson and Stone 2003; Oberdörster et al. 2005a]. Toxicological studies have shown that the chemical and physical properties that influence the fate and toxicity of ultrafine particles may also be relevant to mechanisms influencing biological exposure and response to other nanoscale particles [Duffin et al. 2002; Kreyling et al. 2002; Oberdörster et al. 2002; Semmler et al. 2004; Nel et al. 2006].

Hypothesis 2: Surface area and activity and particle number may be better predictors of potential hazard than mass.

The greater potential hazard may relate to the greater number or surface area of nanoparticles compared with that for the same mass concentration of larger particles [Oberdörster et al. 1992, 1994a,b; Driscoll et al. 1996; Tran et al. 2000; Brown et al. 2001; Peters et al. 1997; Moshammer and Neuberger 2003; Sager et al. 2008]. This hypothesis is based primarily on the pulmonary effects observed in studies of rodents exposed to various types of ultrafine or fine particles (i.e., TiO_2, carbon black, barium sulfate, carbon black, diesel soot, coal fly ash, toner) and in humans exposed to aerosols, including nanoscale particles (e.g., diesel exhaust, welding fumes). These studies indicate that for a given mass of particles, relatively insoluble nanoparticles are more toxic than larger particles of similar chemical composition and surface properties. Studies of fine and ultrafine particles have shown that particles with less reactive surfaces are less toxic [Tran et al. 1999; Duffin et al. 2002]. However, even particles with low inherent toxicity (e.g., TiO_2) have been shown to cause pulmonary inflammation, tissue damage, and fibrosis at sufficiently high particle surface area doses [Oberdörster et al. 1992, 1994a,b; Tran et al. 1999, 2000].

Through engineering, the properties of nanomaterials can be modified. For example, a recent study has shown that the cytotoxicity of water-soluble fullerenes can be reduced by several orders of magnitude by modifying the structure of the fullerene molecules (e.g., by hydroxylation) [Sayes et al. 2004]. These structural modifications were shown to reduce the cytotoxicity by reducing the generation of oxygen radicals—which is a probable mechanism by which cell membrane damage and death occurred in these cell cultures. Increasing the sidewall functionalization of SW-CNT also rendered these nanomaterials less cytotoxic to cells in culture [Sayes et al. 2005]. Cytotoxicity studies with quantum dots have

shown that the type of surface coating can have a significant effect on cell motility and viability [Hoshino et al. 2004; Shiohara et al. 2004; Lovric et al. 2005]. Differences in the phase composition of nanocrystalline structures can influence their cytotoxicity; in a recent study comparing two types of TiO_2 nanoparticles exposed to UV radiation, anatase TiO_2 was more cytotoxic and produced more reactive species than did rutile TiO_2 with similar specific surface area (153 m^2g and 123 m^2g of TiO_2, respectively) [Sayes et al. 2006]. Reactive oxygen species were also associated with the cytotoxicity of TiO_2 nanoparticles to mouse microglia (brain cells) grown in culture [Long et al. 2006]. In contrast, in vitro generation of oxidant species is relatively low in purified SWCNT (contaminating metals removed), yet this material caused progressive interstitial fibrosis in vivo [Shvedova et al. 2004; 2005]. However, recent in vitro studies indicate that purified SWCNTs enhance proliferation and collagen production in fibroblasts [Wang et al. 2008]. Therefore, oxidant generation may not be the only mechanism driving the biological activity of nanomaterials.

The studies of ultrafine particles may provide useful data to develop preliminary hazard or risk assessments and to generate hypotheses for further testing. The studies in cell cultures provide information about the cytotoxic properties of nanomaterials that can guide further research and toxicity testing in whole organisms. More research is needed of the specific particle properties and other factors that influence the toxicity and disease development, including those characteristics that may be most predictive of the potential safety or toxicity of newly engineered nanomaterials.

Very little is known about the safety risks that engineered nanomaterials might pose, beyond some data indicating that they possess certain properties associated with safety hazards in traditional materials. Based upon currently available information, the potential safety concerns most likely would involve catalytic effects or fire and explosion hazards if nanomaterials are found to behave similarly to traditional materials.

6.1 Fire and Explosion Risk

Although insufficient information exists to predict the fire and explosion risk associated with nanoscale powders, **nanoscale combustible material could present a higher risk than a similar quantity of coarser material, given its unique properties** [HSE 2004]. Decreasing the particle size of combustible materials can increase combustion potential and combustion rate, leading to the possibility of relatively inert materials becoming highly reactive in the nanometer size range. Dispersions of combustible nanomaterial in air may present a greater safety risk than dispersions of non-nanomaterials with similar compositions. Some nanomaterials are designed to generate heat through the progression of reactions at the nanoscale. Such materials may present a fire hazard that is unique to engineered nanomaterials. In the case of some metals, explosion risk can increase significantly as particle size decreases.

The greater activity of nanoscale materials forms a basis for research into nanoenergetics. For instance, nanoscale Al/MoO_3 thermites ignite more than 300 times faster than corresponding micrometer-scale material [Granier and Pantoya 2004].

6.2 Risks of Catalytic Reactions

Nanoscale particles and nanostructured porous materials have been used as effective catalysts for increasing the rate of reactions or decreasing the necessary temperature for reactions to occur in liquids and gases. **Depending on their composition and structure, some nanomaterials may initiate catalytic reactions that, based on their chemical composition, would not otherwise be anticipated** [Pritchard 2004].

There are currently no national or international consensus standards on measurement techniques for nanomaterials in the workplace. If the qualitative assessment of a process has identified potential exposure points and leads to the decision to measure nanomaterials, several factors must be kept in mind. Current research indicates that mass and bulk chemistry may be less important than particle size, surface area, and surface chemistry (or activity) for nanostructured materials [Oberdörster et al. 1992, 1994a,b; Duffin et al. 2002]. Research is ongoing into the relative importance of these different exposure metrics, and how to best characterize exposures to nanomaterials in the workplace. In addition, the unique shape and properties of some nanomaterials may pose additional challenges. For example, the techniques used to measure fiber concentrations in the workplace (e.g., phase contrast microscopy) would not be able to detect individual carbon nanotubes with diameters less than 100 nm nor bundles of carbon nanotubes with diameters less than 250 nm [Donaldson et al. 2006]. NIOSH and the National Institute of Standards and Technology (NIST) are collaborating on efforts to develop nanoscale reference materials for exposure assessment. Initial effort is focused on development of TiO_2 reference material.

7.1 Workplace Exposures

While research continues to address questions of nanomaterial toxicity, a number of exposure assessment approaches can be used to help determine worker exposures to airborne nanomaterials. These assessments can be performed using traditional industrial hygiene sampling methods including samplers placed at static locations (area sampling), samples collected in the breathing zone of the worker (personal sampling), or real-time devices or methods that can be personal or static. In general, personal sampling is preferred to ensure an accurate representation of the worker's exposure, whereas area samples (e.g., size-fractionated aerosol samples) and real-time (direct-reading) exposure measurements may be more useful for evaluating the need for improvement of engineering controls and work practices.

Many of the sampling techniques that are available for measuring nanoaerosols vary in complexity but can provide useful information for evaluating occupational exposures with respect to particle size, mass, surface area, number concentration, composition, and surface chemistry. Unfortunately, relatively few of these techniques are readily applicable to routine exposure monitoring. Research is ongoing into developing an analytical strategy for determining both TiO_2 surface area and titanium mass from 37-mm cassette filter samplers. Current measurement techniques are described below along with their applicability for monitoring nanometer aerosols.

For each measurement technique used, it is vital that the key parameters associated with the technique and sampling methodology be recorded when measuring exposure to nanoaerosols. This should include the response range of the instrumentation, whether personal or static measurements are made, and the location of all potential aerosol sources including background aerosols. Comprehensive documentation will facilitate comparison of exposure measurements using different instruments or different exposure metrics and will aid the re-interpretation of historic data as further information is developed on health-appropriate exposure metrics. **Regardless of the metric and method selected for exposure monitoring, it is critical that measurements be taken before production or processing of a nanomaterial to obtain background nanoparticle exposure data.** Measurements made during production or processing can then be evaluated to determine if there has been an increase in particle number concentrations in relation to background measurements and whether that change represents worker exposure to the nanomaterial. Table 7–1 gives a listing of instruments and measurement methods that can be used in the evaluation of engineered nanoparticle exposures.

7.1.1 Size-fractionated aerosol sampling

Studies indicate that particle size plays an important role in determining the potential adverse effects of nanomaterials in the respiratory system: by influencing the physical, chemical, and biological nature of the material; by affecting the surface-area dose of deposited particles; and by enabling deposited particles to more readily translocate to other parts of the body. Animal studies indicate that the toxicity of inhaled nanoparticles is more closely associated with the particle surface area and particle number than with the particle mass concentration when comparing aerosols with different particle size distributions. However, mass concentration measurements may be applicable for evaluating occupational exposure to nanometer aerosols where a good correlation between the surface area of the aerosol and mass concentration can be determined or if toxicity data based on mass dose are available for a specific nanoscale particle associated

Figure 7–1. Examples of different sampling instruments used to measure occupational exposures to nanoparticles including the determination of real-time particle number concentrations and size-fractionated mass concentrations

with a known process (e.g., diesel exhaust particulate).

Aerosol samples can be collected using inhalable, thoracic, or respirable samplers, depending on the region of the respiratory system most susceptible to the inhaled particles. Since prevailing **information suggests that a large fraction of inhaled nanoparticles will deposit in the gas-exchange region of the lungs [ICRP 1994], respirable samplers would be appropriate.** Respirable samplers will also collect a nominal amount of nanoscale particles that can deposit in the upper airways and ultimately be cleared or transported to other parts of the body.

Table 7–1. Summary of instruments and measurement methods used in the evaluation of nanomaterial exposures[*]

Metric	Instrument or method	Remarks
Mass-Direct (total and/ or elemental)	Size Selective Static Sampler	The only instruments offering a cut point around 100 nm are cascade impactors (Berner-type low pressure impactors, or Micro orifice impactors). Allows gravimetric and chemical analysis of samples on stages below 100 nm.
	TEOM (Tapered Element Oscillating Microbalance)	Sensitive real-time monitors such as the TEOM may be useable to measure nanoaerosol mass concentration on-line with a suitable size selective inlet.
	Filter collection and elemental analysis	Filters may be collected with size selective pre-samplers or open face. Elemental analysis (e.g., carbon, metals) for mass determination.
Mass-Indirect (calculation)	ELPI™ (Electrical Low Pressure Impactor)	Real time size-selective (aerodynamic diameter) detection of active surface area concentration giving aerosol size distribution. Mass concentration of aerosols can be calculated when particle charge and density are known or assumed.
	MOUDI (Micro-Orfice Uniform Deposit Impactor)	Real time size-selective (aerodynamic diameter) by cascade impaction.
	DMAS (Differential Mobility Analyzing System)	Real time size-selective (mobility diameter) detection of number concentration, giving aerosol size distribution. Mass concentration of aerosols can be calculated when particle shape and density are known or assumed.

(continued)

See footnotes at end of table.

Table 7–1 (Continued). Summary of instruments and measurement methods used in the evaluation of nanomaterial exposures*

Metric	Instrument or method	Remarks
Number-Direct	CPC (Condensation Particle Counter)	CPCs provide real time number concentration measurements between their particle diameter detection limits. Without a nanoparticle pre-separator they are not specific to the nanometer size range. Some models have diffusion screen to limit top size to 1 μm.
	OPC (Optical Particle Counter)	OPCs provide real time number concentration measurements between their particle diameter detection limits. Particle size diameters begin at 300 nm and may go up to 10,000 nm.
	DMAS and SMPS (Scanning Mobility Particle Sizer)	Real time size-selective (mobility diameter) detection of number concentration giving number-based size distribution.
	Electron Microscopy	Off-line analysis of electron microscope samples can provide information on size-specific aerosol number concentration.
Number-Indirect	ELPI™ and MOUDI	Real time size-selective (aerodynamic diameter) detection of active surface-area concentration giving aerosol size distribution. Data may be interpreted in terms of number concentration. Size-selective samples may be further analyzed off-line.
Surface Area-Direct	Diffusion Charger	Real-time measurement of aerosol active surface-area. Active surface-area does not scale directly with geometric surface-area above 100 nm. Note that not all commercially available diffusion chargers have a response that scales with particle active surface-area below 100 nm. Diffusion chargers are only specific to nanoparticles if used with appropriate inlet pre-separator.
	ELPI™ and MOUDI	Real-time size-selective (aerodynamic diameter) detection of active surface-area concentration. Active surface-area does not scale directly with geometric surface-area above 100 nm.

(continued)

See footnotes at end of table.

Table 7–1 (Continued). Summary of instruments and measurement methods used in the evaluation of nanomaterial exposures*

Metric	Instrument or method	Remarks
Surface Area-Direct (continued)	Electron Microscopy	Off-line analysis of electron microscope samples (previously collected on filters or other media) can provide information on particle surface-area with respect to size. TEM analysis provides direct information on the projected area of collected particles which may be related to geometric area for some particles shapes.
Surface Area-Indirect (calculation)	DMAS and SMPS	Real time size-selective (mobility diameter) detection of number concentration. Data may be interpreted in terms of aerosol surface-area under certain circumstances. For instance, the mobility diameter of open agglomerates has been shown to correlate with projected surface area.
	DMAS and ELPI™ used in parallel	Differences in measured aerodynamic and mobility can be used to infer particle fractal dimension which can be further used to estimate surface-area.

*Adapted from ISO/TR 12885

Note: Inherent to all air sampling instruments in this table is the fact that they cannot discriminate the nanoaerosol of interest from other airborne particles. Also, there is a general lack of validation regarding the response of these air sampling instruments to the full spectrum of nanoparticles that may be found in the workplace, including varieties of primary particles, agglomerates or aggregates, and other physical and chemical forms. A suite of nanoparticle reference materials are required to perform the needed validations.

Respirable samplers allow mass-based exposure measurements to be made using gravimetric and/or chemical analysis [NIOSH 1994]. However, they do not provide information on aerosol number, size, or surface-area concentration, unless the relationship between different exposure metrics for the aerosol (e.g., density, particle shape) has been previously characterized. Currently, no commercially available personal samplers are designed to measure the particle number, surface area, or mass concentration of nanoaerosols. However, several methods are available that can be used to estimate surface area, number, or mass concentration for particles smaller than 100 nm.

The use of conventional impactor samplers to assess nanoparticle exposure is limited since the impaction collection efficiencies are 200–300 nm. Low-pressure cascade impactors that can measure particles to 50 nm and larger may be used for static sampling since their size and complexity preclude their use as personal samplers [Marple et al. 2001; Hinds 1999]. A personal cascade impactor is available with a lower aerosol cut point of

250 nm [Misra et al. 2002], allowing an approximation of nanoscale particle mass concentration in the worker's breathing zone. For each method, the detection limits are on the order of a few micrograms of material on a filter or collection substrate [Vaughan et al. 1989]. Cascade impactor exposure data gathered from worksites where nanomaterials are being processed or handled can be used to make assessments as to the efficacy of exposure control measures.

7.1.2 Real-time aerosol sampling

The real-time (direct-reading) measurement of nanometer aerosol concentrations is limited by the sensitivity of the instrument to detect small particles. Many real-time aerosol mass monitors used in the workplace rely on light scattering from groups of particles (photometers). This methodology is generally insensitive to particles smaller than 100 nm [Hinds 1999]. Optical instruments that size individual particles and convert the measured distribution to a mass concentration are similarly limited to particles larger than 100 nm. Quantitative information gained by optical particle counters may also be limited by relatively poor counting efficiencies at smaller particle diameters (i.e., less than 500 nm). These instruments are capable of operating within certain concentration ranges that, when exceeded, affect the count or mass determination efficiencies due to coincidence errors at the detector. Similarly, the response of optical particle counters may be material-dependent according to the refractive index of the particle. The Scanning Mobility Particle Sizer (SMPS) is widely used as a research tool for characterizing nanoscale aerosols although its applicability for use in the workplace may be limited

because of its size, cost, and the inclusion of a radioactive source. Additionally, the SMPS may take 2–3 minutes to scan an entire size distribution; thus, it may be of limited use in workplaces with highly variable aerosol size distributions, such as those close to a strong particle source. Fast (less than 1 second), mobility-based, particle-sizing instruments are now available commercially; however, having fewer channels, they lack the finer sizing resolution of the SMPS. The Electrical Low Pressure Impactor (ELPI) is an alternative instrument that combines diffusion charging and a cascade impactor with real-time (less than 1 second) aerosol charge measurements providing aerosol size distributions by aerodynamic diameter [Keskinen et al. 1992].

7.1.3 Surface-area measurements

Relatively few techniques exist to monitor exposures with respect to aerosol surface area. Particle surface is composed of internal surface area attributable to pores (cavities more deep than wide), external surface area due to roughness (cavities more wide than deep), and total surface area (the accessible area of all real particle surfaces). A standard gas adsorption technique (i.e., BET) is used to measure the total surface area of powders and can be adapted to measure the specific surface area (surface area per unit mass) of engineered nanomaterials [Brunauer et al. 1938]. However, surface-area analysis by gas adsorption requires relatively large quantities of material, is not element specific, and must be performed in a laboratory.

The first instrument designed to measure aerosol surface area was the epiphaniometer [Baltensperger et al. 1988]. This device measures the Fuchs, or active surface area, of the aerosols by measuring the attachment rate of radioactive ions. For aerosols less than approximately 100 nm in size, measurement of the Fuchs surface area is probably a good indicator of external surface area (or geometric surface area). However, for aerosols greater than approximately 1 μm, the relationship with geometric particle surface area is lost [Fuchs 1964]. Measurements of active surface area are generally insensitive to particle porosity. The epiphaniometer is not well suited to widespread use in the workplace because of the inclusion of a radioactive source and the lack of effective temporal resolution.

This same measurement principle can be applied with the use of a portable aerosol diffusion charger. Studies have shown that these devices provide a good estimate of aerosol external surface area when airborne particles are smaller than 100 nm in diameter. For larger particles, diffusion chargers underestimate aerosol surface area. However, further research is needed to evaluate the degree of underestimation. Extensive field evaluations of commercial instruments are yet to be reported. However, laboratory evaluations with monodisperse silver particles have shown that two commercially available diffusion chargers can provide good measurement data on aerosol external surface area for particles smaller than 100 nm in diameter but underestimate the aerosol surface area for particles larger than 100 nm in diameter [Ku and Maynard 2005, 2006].

7.1.4 Particle number concentration measurement

Particle number concentration has been associated with adverse responses to air pollution in some human studies [Timonen et al. 2004; Ruckerl et al. 2005], while in toxicologic studies, particle surface area has

generally been shown to be a better predictor than either particle number, mass, or volume concentration alone [Oberdörster and Yu 1990; Tran et al. 1999; Duffin et al. 2002]. A two-variable dose metric of particle size and volume has been shown to be the best predictor of lung cancer in rats from various types of particles [Borm et al. 2004; Pott and Roller 2005]. This illustrates some of the complexity of interpreting existing data on particle dose metric and response. While adverse health effects appear to be more closely related with particle surface area, the number of particles depositing in the respiratory tract or other organ systems may also play an important role.

Aerosol particle number concentration can be measured relatively easily using Condensation Particle Counters (CPCs). These are available as hand-held static instruments, and they are generally sensitive to particles greater than 10–20 nm in diameter. Condensation Particle Counters designed for the workplace do not have discrete size-selective inputs, and so they are typically sensitive to particles less than 1 μm in diameter. Commercial size-selective inlets are not available to restrict CPCs to the nanoparticle size range; however, the technology exists to construct size-selective inlets based on particle mobility or possibly on inertial pre-separation. An alternative approach to estimating nanoparticle number concentrations using a CPC is to use the instrument in parallel with an optical particle counter (OPC). The difference in particle count between the instruments will provide an indication of particle number concentration between the lower CPC detectable particle diameter and the lower OPC particle diameter (typically 300–500 nm).

A critical issue when characterizing exposure using particle number concentration is selectivity. **Nanoscale particles are ubiquitous in many workplaces**, from sources such as combustion, vehicle emissions, and infiltration of outside air. Particle counters are generally insensitive to particle source or composition **making it difficult to differentiate between incidental and process-related nanoparticles using number concentration alone.** In a study of aerosol exposures during a carbon black bagging process, Kuhlbusch et al. [2004] found that peaks in number concentration measurements were associated with emissions from fork lift trucks and gas burners in the vicinity, rather than with the process itself. In a similar manner, during an ultrafine particle mapping exercise in an automotive facility, Peters et al. [2006] found that direct gas-fired heating systems systematically produced high particle number concentrations throughout the facility when the heating system was in operation. Through follow up measurements, Heitbrink et al. [2007] found a high proportion of ultrafine particles produced from these burners, yet little if any mass was associated with their emissions. Other non-process ultrafine sources were identified in an adjacent foundry [Evans et al. 2008]. Together with roof mounted gas-fired heating units, additional sources included cigarette-smoking and the exhaust from a propane fueled sweeper vehicle, with the latter contributing a large fraction of the ultrafine particles. Although these issues are not unique to particle number concentration measurements, orders of magnitude difference can exist in particle number concentrations depending on concomitant sources of particle emissions.

Although using nanoparticle number concentration as an exposure measurement may not be consistent with exposure metrics being used in animal toxicity studies, **such measurements may be useful for identifying nanoscale particle emissions and determining the efficacy of control measures.** Portable CPCs are capable of measuring localized particle concentrations allowing the assessment of particle releases occurring at various processes and job tasks [Brouwer et al. 2004].

7.1.5 Surface-area estimation

Information about the relationship between different measurement metrics can be used for approximating particle surface area. If the size distribution of an aerosol remains consistent, the relationship between particle number, surface area, and mass metrics will be constant. In particular, mass concentration measurements can be used for deriving surface-area concentrations, assuming the constant of proportionality is known. This constant is the specific surface area (surface to mass ratio).

Size distribution measurements may be obtained through the collection of filter samples and analysis by transmission electron microscopy to estimate particle surface area. If the measurements are weighted by particle number, information about particle geometry will be needed to estimate the surface area of particles with a given diameter. If the measurements are weighted by mass, additional information about particle density will be required. Estimates of particle-specific surface area from geometric relation with external particle dimensions depends upon the morphology regime of the material of interest and is only appropriate for smooth, regularly shaped, compact particles [Stefaniak et al. 2003; Weibel et al. 2005]. For example, Weibel et al. [2005] report that estimates of ultrafine TiO_2 surface area determined using a geometric relationship with the physical particle size (using TEM) were 50% lower than measured using nitrogen gas adsorption.

If the airborne aerosol has a lognormal size distribution, particle surface-area concentration can be derived using three independent measurements. An approach has been proposed using three simultaneous measurements of the aerosol that included mass concentration, number concentration, and charge [Woo et al. 2001]. With knowledge of the response function of each instrument, minimization techniques can be used to estimate the parameters of the lognormal distribution leading to the three measurements used in estimating the particle surface area.

An alternative approach has been proposed whereby independent measurements of particle number and mass concentration are made, and the surface area is estimated by assuming the geometric standard deviation of the (assumed) lognormal distribution [Maynard 2003]. This method has the advantage of simplicity by relying on portable instruments that can be used in the workplace. Theoretical calculations have shown that estimates may be up to a factor of 10 different from the actual particle surface area, particularly when the aerosol has a bimodal distribution. Field measurements indicate that estimates are within a factor of 3 of the active surface area, particularly at higher concentrations. In workplace environments, particle surface-area concentrations can be expected to span up to 5 orders of magnitude; thus, surface-area estimates may be suited for initial or preliminary appraisals of occupational exposure concentrations.

Although such estimation methods are unlikely to become a long-term alternative to more accurate methods, they may provide a viable interim approach to estimating the surface area of nanoscale particles in the absence of precise measurement data. Additional research is needed on comparing methods used for estimating particle surface area with a more accurate particle surface-area-measurement method. NIOSH is conducting research in this area and will communicate results as they become available.

7.1.6 Particle number concentration mapping

To better understand particle sources and contaminant migration, some investigators have adopted an aerosol mapping technique, which integrated measurements of respirable mass, ultrafine particle number, and active surface-area concentrations in automotive manufacturing facilities [Peters et al. 2006; Heitbrink et al. 2007, 2008; Evans et al. 2008]. The process relies on portable aerosol sampling instrumentation for simultaneous measurements at predetermined positions throughout a facility. The technique is somewhat measurement-intensive but is useful for locating contaminant sources and determining the extent of contaminant migration. Leaks and other less obvious particle sources have been identified in this way and the procedure provides a powerful tool for facility staff to target their contaminant control approaches most effectively. This technique relies on successive measurements at various locations, making facilities with continuous processes or those likely to achieve steady state particle number concentrations most appropriate for this approach. The approach is less successful for facilities with batch processes or those likely to experience rapid concentration changes as, depending on where in the measurement cycle the release occurs, it may be overlooked. A high degree of variability between mapping events is expected in

7.2 Sampling Strategy

Currently, there is not one sampling method that can be used to characterize exposure to nanoscale aerosols. Therefore, any attempt to characterize workplace exposure to nanomaterials must involve a multifaceted approach incorporating many of the sampling techniques mentioned above. Brouwer et al. [2004] recommend that all relevant characteristics of nanomaterial exposure be measured, and a sampling strategy similar to theirs would provide a reasonable approach to characterizing workplace exposure. NIOSH has developed the Nanoparticle Emission Assessment Technique (NEAT) to qualitatively determine the release of engineered nanomaterials in the workplace (see Appendix). This approach may be helpful to others for the initial evaluation of workplaces where engineered nanomaterials are manufactured or used. If material release is found and if resources allow, then a more comprehensive and quantitative approach may be adopted [Methner et al. 2007].

The first step to characterizing workplace exposures would involve identifying the source of nanomaterial emissions. A CPC used in parallel with an OPC provides acceptable capability for this purpose. **It is critical to determine ambient or background particle counts before measuring particle counts during the manufacturing, processing, or handling of engineered nanomaterials.** However, investigators need to be aware that background nanoscale particle counts can vary both spatially and temporally depending on the unique conditions of the workplace. Subtraction of background nanoscale particle counts will be most challenging in these situations. In cases where nanomaterial handling or processing operations contribute only small elevations in particle counts, it may not be possible to adequately characterize these increases, particularly if the background particle count is relatively high.

If nanomaterials are detected in the process area at elevated concentrations relative to background particle number concentrations, then a pair of filter-based, area air samples should be collected for particle analysis via transmission electron microscopy (TEM) and for determining mass concentration. Transmission electron microscopy can provide an estimate of the particle size distribution and, if equipped with an energy dispersive X-ray analyzer (EDS), a determination of elemental composition

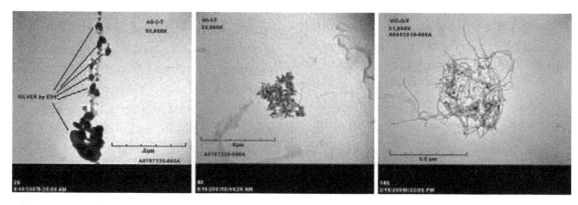

Figure 7–2. Photomicrographs of airborne engineered nanomaterials (airborne samples of engineered nanoparticles of silver, nickel, and MWCNT analyzed by TEM and EDS)

can be made to identify the nanomaterial (see Figure 7–2).

Analysis of filters for mass determination of air contaminants of interest can help identify the source of the particles. Standard analytical chemical methodologies (e.g., NMAM 5040 for carbon, NMAM 7303 for metals) should be employed [NIOSH 1994].

The combination of particle counters and samples for chemical analysis allows for an assessment of worker exposure to nanomaterials (see Figure 7–3) and the characterization of the important aerosol metrics. However, since this approach relies primarily on static or area sampling, some uncertainty will exist in estimating worker exposures.

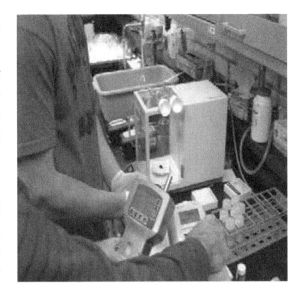

Figure 7–3. Combined use of the OPC, CPC, and two filter samples to determine the presence of nanomaterials

8 Guidelines for Working with Engineered Nanomaterials

Engineered nanomaterials are diverse in their physical, chemical, and biological nature. The processes used in research, material development, production, and use or introduction of nanomaterials have the potential to vary greatly. **Until further information on the possible health risks and extent of occupational exposure to nanomaterials becomes available, interim protective measures should be developed and implemented.** These measures should focus on the development of engineering controls and safe working practices tailored to the specific processes and materials where workers might be exposed. Hazard information that is available about common materials being manufactured in the nanoscale range (e.g., TiO_2, beryllium) should be considered as a starting point in developing appropriate controls and work practices.

The following recommendations are designed to aid in the assessment and control of workplace exposures to engineered nanomaterials. Using a hazard-based approach to evaluate exposures and for developing precautionary measures is consistent with good occupational safety and health practices [The Royal Society and The Royal Academy of Engineering 2004; Schulte et al. 2008].

8.1 Potential for Occupational Exposure

Few workplace measurement data exist on airborne exposure to nanomaterials that are purposely produced and not incidental to an industrial process. In general, it is likely that processes generating nanomaterials in the gas phase (after removal of the nanomaterial from an enclosed generation system), or using or producing nanomaterials as powders or slurries/suspensions/solutions (i.e., in liquid media), pose the greatest risk for releasing nanoparticles. In addition, **maintenance on production systems (including cleaning and disposing of materials from dust collection systems) is likely to result in exposure to nanoparticles if deposited nanomaterials are disturbed.** Exposures associated with waste streams containing nanomaterials may also occur.

The magnitude of exposure to nanomaterials when working with nanopowders depends on the likelihood of particles being released from the powders during handling. NIOSH is actively conducting research to quantitatively determine how various nanomaterials are dispersed in the workplace. Studies on exposure to SWCNTs and MWCNTs have indicated that the raw material may release visible particles into the air when handled, that the particle size of the agglomerate can be a few millimeters in diameter, and that the release rate of inhalable and respirable particles is relatively low (on a mass or number basis) compared with other nanopowders. Maynard et al. [2004] reported concentrations of respirable dust from 0.007 to 0.053 mg/m^3 when energy was applied (vortexing) to bulk SWCNT for approximately 30 minutes. Similar findings were reported by Han et al. [2008] at a laboratory producing MWCNTs in which exposure concentrations as high as 0.4 mg/m^3 were observed prior to the implementation of exposure controls. In a health hazard evaluation conducted by NIOSH at a university-based research laboratory the potential release of airborne carbon nanotubes (CNFs) was observed at various processes [Methner et al. 2007]. General area exposure measurements indicated slight increases in airborne particle number and mass concentrations relative to background measurements during the transfer of CNFs prior to weighing and mixing, and during wet saw cutting of a composite material. Since data are lacking on the generation of inhalable/respirable particles during the production and use of engineered nanomaterials, further research is required to determine exposures under various conditions. NIOSH researchers are conducting both laboratory and field-based evaluations in order to address some of these knowledge gaps.

Devices comprised of nanostructures, such as integrated circuits, pose a minimal risk of exposure to nanomaterials during handling. However, some of the processes used in their production may lead to exposure to nanomaterials (e.g., exposure to commercial polishing compounds that contain nanoscale particles, exposure to nanoscale particles that are inadvertently dispersed or created during the manufacturing and handling processes). Likewise, large-scale components formed from nanocomposites will most likely not present significant exposure potential. However, if such materials are used or handled in such a manner that can generate nanoparticles (e.g., cutting, grinding) or undergo degradation processes that lead to the release of nanostructured material, then exposure may occur by the inhalation, ingestion, and/or dermal penetration of these particles.

8.2 Factors Affecting Exposure to Nanomaterials

Factors affecting exposure to engineered nanomaterials include the amount of material being used and whether the material can be easily dispersed (in the case of a powder) or form airborne sprays or droplets (in the case of suspensions). The degree of containment and duration of use will also influence exposure. In the case of airborne material, particle or droplet size will determine whether the material can enter the respiratory tract and where it is most likely to deposit. Respirable particles are those capable of depositing in the alveolar (gas exchange) region of the lungs, which includes particles smaller than approximately 10 μm in diameter [Lippmann 1977; ICRP 1994; ISO 1995]. The proportion of inhaled nanoparticles likely to deposit in any region of the human respiratory tract is approximately 30%–90% depending on factors such as breathing rate and particle size. Up to 50% of nanoparticles in the 10–100 nm size range may deposit in the alveolar region, while nanoparticles smaller than 10 nm are more likely to deposit in the head and thoracic regions [ICRP 1994]. **The mass deposition fraction of inhaled nanoparticles in the gas-exchange region of the lungs is greater than that for larger respirable particles.**

At present there is insufficient information to predict all of the situations and workplace scenarios that are likely to lead to exposure to nanomaterials. However, there are some workplace factors that can increase the potential for exposure:

- working with nanomaterials in liquid media without adequate protection (e.g., gloves)

- working with nanomaterials in liquid during pouring or mixing operations or where a high degree of agitation is involved

- generating nanomaterials in the gas phase in non-enclosed systems

- handling (e.g., weighing, blending, spraying) powders of nanostructured materials

- maintenance on equipment and processes used to produce or fabricate nanomaterials

- cleaning up spills or waste material

- cleaning dust collection systems used to capture nanoparticles

- machining, sanding, drilling of nanomaterials, or other mechanical disruptions of nanomaterials can potentially lead to the aerosolization of nanoparticles.

8.3 Elements of a Risk Management Program

Given the limited information about the health risks associated with occupational exposure to engineered nanomaterials, appropriate steps should be taken to minimize the risk of worker exposure through the implementation of a risk management program [Schulte et al. 2008]. Risk management programs for nanomaterials should be seen as an integral part of an overall occupational safety and health program for any company or workplace producing or using nanomaterials or nanoenabled products. A critical element of the program should be the capability to anticipate new and emerging risks (hazard determination) and whether they are linked to changes in the manufacturing process, equipment, or the introduction of new materials. This will require an ongoing assessment of the potential risks to workers (risk evaluation) through the systematic collection of job and product information so that determinations can be made regarding scenarios (e.g., laboratory research, production and manufacture, nanoenabled product use) that place the worker in contact with nanomaterials (see Figure 8–1). This assessment should be an ongoing cyclic process that provides feedback on potential sources of exposure and solutions taken to correct those problems. For example, operations and job tasks that have the potential to aerosolize nanomaterials (e.g., handling dry powders, spray applications) deserve more attention and more stringent controls than those where the nanomaterials are imbedded in solid or liquid matrices. Elements of the risk management program should include guidelines for installing and evaluating engineering controls (e.g., exhaust ventilation, dust collection systems), the education and training of workers in the proper handling of nanomaterials (e.g., good work practices), and the selection and use of personal protective equipment (e.g., clothing, gloves, respirators).

When controlling potential exposures within a workplace, NIOSH has recommended a hierarchical approach to reduce worker exposures (see Table 8–1) [NIOSH 1990]. The philosophical basis for the hierarchy of controls is to eliminate the hazard when possible (i.e., substitute with a less hazardous material) or, if not feasible, control the hazard at or as close to the source as possible.

8.3.1 Engineering Controls

If the potential hazard can not be eliminated or substituted with a less hazardous or non-hazardous substance, then engineering controls should be installed and tailored to the process or job task. The type of engineering control used should take into

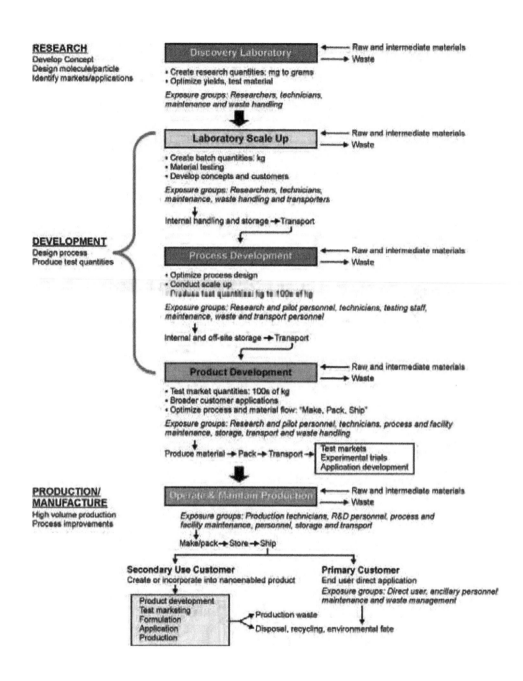

Figure 8–1. Workplaces with potential for occupational exposure to engineered nanomaterials. The figure illustrates the life cycle of nanomaterials from laboratory research development through product development, use, and disposal. Each step of the life cycle represents opportunities for potential worker exposure to nanomaterials. Adapted from Schulte et al. 2008a.

Table 8–1. Hierarchy of exposure controls*

Control method	Process, equipment, or job task
1. Elimination	Change design to eliminate hazard
2. Substitution	Replace a high hazard for a low hazard
3. Engineering	Isolation/enclosure, ventilation (local, general)
4. Administrative	Procedures, policies, shift design
5. Personal protective equipment	Respirators, clothing, gloves, goggles, ear plugs

*Control methods are typically implemented in this order to limit worker exposures to an acceptable concentration (e.g., occupational exposure limit or other pre-established limit).
Sources: Plog et al. 2002; NIOSH 1990.

account information on the potential hazardous properties of the precursor materials and intermediates as well as those of the resulting nanomaterial. In light of current scientific knowledge about the generation, transport, and capture of aerosols [Seinfeld and Pandis 1998; Hinds 1999], airborne exposure to nanomaterials can most likely be controlled at most processes and job tasks using a wide variety of engineering control techniques similar to those used in reducing exposures to general aerosols [Ratherman 1996; Burton 1997].

Engineering control techniques such as source enclosure (i.e., isolating the generation source from the worker) and local exhaust ventilation systems should be effective for capturing airborne nanomaterials, based on what is known of nanomaterial motion and behavior in air (see Figure 8–2). The quantity of the bulk nanomaterial that is synthesized or handled in the manufacture of a product will significantly influence the selection of the exposure controls.

Other factors that influence selection of engineering controls include the physical form of the nanomaterial and task duration and frequency. For instance, working with nanomaterial in the slurry form in low quantities would require a less rigorous control system than those that would be required for large quantities of nanomaterials in a free or fine powder form (see Figure 8–3). Unless cutting or grinding occurs, nanomaterials that are not in free form (encapsulated in a solid, nanocomposites, and surface coated materials) typically wouldn't require engineering controls.

Handling research quantities typically occurs in laboratories with ventilation controls. Since quantities are small, local containment and control can be applied, such as low-flow vented work stations and small glove box chambers. However, as quantities are increased, care must be taken to reduce the amount of nanomaterial that is released from the process equipment and to prevent the migration of nanomaterials into adjacent rooms or areas. For example, the installation of local exhaust ventilation at a

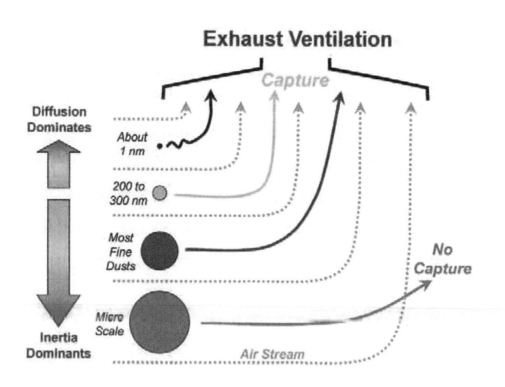

Figure 8–2. Exposure control of particles (illustration of how particle diameter-related diffusion and inertia influence particle capture efficiency in a ventilation system). Particles with a diameter of 200–300 nm have minimal diffusion and inertial properties and are easily transported by moving air and captured. Particle motion by diffusion increasingly dominates as particle diameter decreases below 200 nm. The inertial behavior of larger particles, especially those ejected from energetic processes such as grinding, increases significantly with particle diameter, enabling them to cross the streamlines of moving air and avoid capture. Adapted from Schulte et al. 2008a.

reactor used to make nanoscale engineered metal oxides and metals was found to reduce nanoparticle exposures by 96% (mean particle number concentration) [Methner 2008]. The use of exhaust ventilation systems should be designed, tested, and maintained using approaches recommended by the American Conference of Governmental Industrial Hygienists [ACGIH 2001].

A secondary but nonetheless important issue concerning the control of nanoparticle emissions is that of product loss. The properties of nanomaterials, along with the

unique methods that may be employed for producing them, may mean that traditional exhaust ventilation may be more energetic than necessary for removing incidentally released nanoscale particles. For this reason, engineering controls need to be applied judiciously to ensure protection of workers without compromising production.

8.3.2 Dust collection efficiency of filters

Current knowledge indicates that a well-designed exhaust ventilation system with a

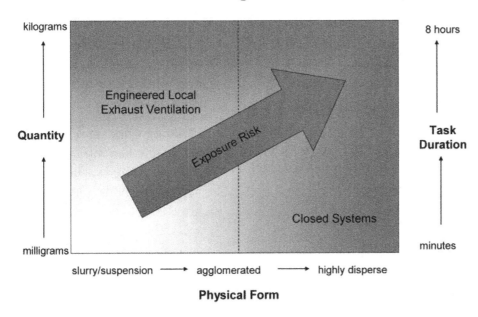

Figure 8–3. Factors influencing control selection. Several factors influence the selection of exposure controls for nanomaterials including quantity of nanomaterial handled or produced, physical form, and task duration. As each one of theses variables increase, exposure risk becomes greater as does the need for more efficient exposure control measures.

HEPA filter should effectively remove nanoparticles [Hinds 1999]. Limited studies have reported the efficacy of filter media typically found in control systems (including respirators) in capturing nanoparticles. The dearth of data on filtration performance against nanoparticles (in particular nanoparticles smaller than 20 nm) is primarily due to the challenges in generating and quantifying particles in those size ranges. Despite these limitations, the results of some studies [Van Osdell et al. 1990] using different filter media challenged with monodispersed aerosols (silver 4–10 nm and dioctylphthalate 32–420 nm) were in agreement with classical single-fiber theory showing an increase in filtration efficacy for smaller size particles. No evidence for particle thermal rebound was

observed. Similar results have been recently reported by Kim et al. [2007] using different filter media challenged with particles ranging in size from 2.5–20 nm, indicating that other filter medias—including those used in air purifying respirators—would behave similarly.

If HEPA filters are used in the dust collection system, they should be coupled with well-designed filter housings. If the filter is improperly seated within the housing, nanoparticles have the potential to bypass the filter, leading to filter efficiencies much less than predicted [NIOSH 2003].

8.3.3 Work practices

An integral step in establishing good work practices is having knowledge of the potential hazards in the workplace and developing formal procedures that describe actions to be taken to ensure the protection of workers. Incorporated in these procedures should be guidelines for good work practices intended to minimize worker exposure to nanomaterials and other potentially hazardous chemicals. Management should systemically review and update these procedures. Actions taken to resolve and/or improve workplace conditions should be routinely conveyed by management to workers.

Good practices for management

- Educating workers on the safe handling of engineered nano-objects or nano-object-containing materials to minimize the likelihood of inhalation exposure and skin contact.

- Providing information, as needed, on the hazardous properties of the precursor materials and those of the resulting nanomaterials product with instruction on measures to prevent exposure.

- Encouraging workers to use hand-washing facilities before eating, smoking, or leaving the worksite.

- Providing additional control measures (e.g., use of a buffer area, decontamination facilities for workers if warranted by the hazard) to ensure that engineered nanomaterials are not transported outside the work area [US DOE 2007].

- Providing facilities for showering and changing clothes to prevent the inadvertent contamination of other areas (including take-home) caused by the transfer of nanomaterials on clothing and skin.

Good practices for workers

- Avoiding handling nanomaterials in the open air in a *'free particle'* state.

- Storing dispersible nanomaterials, whether suspended in liquids or in a dry particle form in closed (tightly sealed) containers whenever possible.

- Cleaning work areas at the end of each work shift, at a minimum, using either a HEPA-filtered vacuum cleaner or wet wiping methods. Dry sweeping or air hoses should not be used to clean work areas. Cleanup should be conducted in a manner that prevents worker contact with wastes. Disposal of all waste material should comply with all applicable Federal, State, and local regulations.

- Avoiding storing and consuming food or beverages in workplaces where nanomaterials are handled.

8.3.4 Personal protective clothing

Currently, there are no generally acceptable guidelines available based on scientific data for the selection of protective clothing or other apparel against exposure to nanomaterials. This is due in part to minimal data being available on the efficacy of existing protective clothing, including gloves. In any case, although nanoparticles may penetrate the epidermis, there has been little evidence to suggest that penetration leads to disease; and no dermal exposure standards have been proposed. However, based on a recent survey of nanotechnology workplaces [ICON 2006], 84% of employers recommended personal protective equipment and clothing for employees working with nanomaterials. These recommendations were generally based on conventional occupational hygiene practices but also varied with the size of the company, the type of nanomaterials being handled, and the commercial sector. While some guidelines on the use of protective clothing and gloves have been developed by organizations for use in their own laboratories (e.g., US DOE 2007) or countries (e.g., British Standards Institute BSI 2008) or by consensus standards development organizations (e.g., ASTM, 2007), these are generally based upon good industrial hygiene practices rather than scientific data specific to nanomaterials.

A challenge to making appropriate recommendations for dermal protection against nanoparticles is the need to strike a balance between comfort and protection. Garments that provide the highest level of protection (e.g., an impermeable Level A suit) are also the least comfortable to wear for long periods of time, while garments that are probably the least protective (e.g., thin cotton lab coat) are the most breathable and comfortable for employees to wear. The two primary routes of exposure to particulates for workers using protective clothing are direct penetration through the materials and leakage through gaps, seams, defects, and interface and closure areas [Schneider et al. 1999, 2000]. The relative contributions from these two inward leakage sources are not well-understood. NIOSH has an active research program designed to assess the efficacy of barrier materials and ensembles for protection against particulate hazards, including nanoparticles.

The lack of available data is further complicated by the limitations and difficulties of current test methods, which fall into two basic categories: penetration tests on material swatches to determine barrier efficiency and system-level aerosol testing to determine product ensemble integrity. The former are usually bench-scale testing methods, while the latter require an exposure chamber that is large enough for at least one human test subject or mannequin. Chamber design requirements for system level aerosol testing have been reviewed by Gao et al. [2007]. Little scientific data exists, but some systems level test methods are available. ISO standard method 13982 [ISO 2004a] and EN standard method 943 [CEN 2002] specify the use of sodium chloride (NaCl) with a mass median aerodynamic diameter (MMAD) of 0.6 μm to determine the barrier efficiency of protective clothing against aerosols of dry, fine dusts. The standard method issued by National Fire Protection Association [2007] is a method that is not dependent on filtration-based approaches. Penetration of fluorophore-impregnated silica particles with a MMAD of 2.5 μm and a geometric standard deviation of 2.6 are qualitatively visualized by black light that causes the fluorescent glow of the challenge aerosol particles. Note that the polydisperse particle challenges used in these methods include a large number of nanoscale particles when measured by count rather than by mass.

Particle penetration test methods can be further categorized into those that are analogous to the process used in respirator filter testing and those that are not dependent on filtration-based approaches. Test methods that involve measuring aerosol concentrations using a sampling flow rate do not mimic in-situ situations because the skin does not "breathe." Standardized methodology is needed that is not dependent on filtration-based approaches for examining the overall barrier-effectiveness of the full protective clothing ensemble for different materials to particulate hazards. In this respect, NIOSH has presented preliminary results [Wang and Gao 2007] on development of a magnetic passive aerosol sampler for more accurate determination of particle penetration through protective clothing. NIOSH is conducting research in this area and will communicate results as they become available.

The bulk of the penetration data available on clothing has been done with filtration based testing. One study found that penetration levels of 30–2,000-nm-sized potassium chloride particles through an unidentified military garment ranged from about 20%–60%, with the maximum penetration occurring in the range of 100–400 nm [Hofacre 2006]. Another group of researchers studied the barrier efficiency of 10 unidentified fabric samples (woven, non-woven, and laminated fabrics) using 477-nm-sized latex spheres at a flow rate of 1.8 cm/second [Shavlev et al. 2000]. Particle penetration measurements ranged from 0%–54%, with three of the fabrics exhibiting a measurable pressure drop and having penetration levels less than 1%. In general, these findings suggest that increased external air pressure (e.g., from wind) results in increased particle penetrations. Thus, only impermeable barrier materials are likely to provide complete barrier protection against aerosol penetration. Body movement (i.e., bellows effect) can also impact penetration [Bergman et al. 1989]. NIOSH will theoretically and empirically investigate wind-driven nanoparticle penetration through protective clothing in an attempt to obtain a predictive model based upon single-fiber theory. Results will be communicated as they become available.

Another widely used test method incorporates testing with nanoscale particles in solution, and therefore also provides some indication of the effectiveness of protective clothing to nanoparticles. ASTM standard F1671–03 [ASTM 2003] and ISO standard 16604 [ISO 2004b] specify the use of a 27-nm bacteriophage to evaluate the resistance of

materials used in protective clothing from the penetration of blood-borne pathogens. One study [Edlich et al. 1999] evaluated the integrity of powder-free examination gloves and found that no bacteriophage penetration was detected for powder-free nitrile gloves, powder-free latex gloves, nor polyvinyl chloride synthetic gloves.

Based upon the uncertainty of the health effects of dermal exposure to nanoparticles, it is prudent to consider using protective equipment, clothing, and gloves to minimize dermal exposure, with particular attention given to preventing exposure of nanomaterials to abraded or lacerated skin. Until scientific data exist specific to the performance of protective clothing and gloves against nanomaterials, current industrial hygiene best practices should be followed.

8.3.5 Respirators

The use of respirators is often required when engineering and administrative controls do not adequately keep worker exposures to an airborne contaminant below a regulatory limit or an internal control target. Currently, there are no specific exposure limits in the United States for airborne exposures to engineered nanomaterials although occupational exposure limits and guidelines exist for airborne particles of similar chemical composition regardless of particle size. Current scientific evidence indicates that nanoparticles may be more biologically reactive than larger particles of similar chemical composition and thus may pose a greater health risk when inhaled. In determining the need for respirators, it would therefore be prudent to consider current exposure limits or guidelines (e.g., OSHA PELs, NIOSH RELs, ACGIH TLVs) for larger particles of similar composition, existing toxicologic data on the specific nanoparticle, and the likelihood of worker exposure (e.g., airborne concentration, time exposed, job task).

The decision to institute respiratory protection should be based on a combination of professional judgment and the results of the hazard assessment and risk management practices recommended in this document. The effectiveness of administrative, work-practice, and engineering controls can be evaluated using the measurement techniques described in Chapter 7 Exposure Assessments and Characterization. If worker exposure to airborne nanomaterials remains a concern after instituting control measures, the use of respirators can provide further worker protection. Several classes of respirators exist that can provide different levels of protection when properly fit tested on the worker. Table 8–2 lists various types of particulate respirators that can be used; information is also provided on the level of exposure reduction that can be expected along with the advantages and disadvantages of each respirator type. To assist respirator users, NIOSH has published the document *NIOSH Respirator Selection Logic (RSL)* that provides a process that respirator program administrators can use to select appropriate respirators [NIOSH 2004] (see www.cdc.gov/niosh/docs/2005-100/default.html). As new toxicity data for individual nanomaterials become available, NIOSH will review the data and make recommendations for respirator protection.

When respirators are required for use in the workplace, the Occupational Safety and Health Administration (OSHA) respiratory protection standard [29 CFR 1910.134] requires that a respiratory program be established that includes the following program elements: (*1*) an evaluation of the worker's ability to perform the work while wearing a respirator, (*2*) regular training of personnel, (*3*) periodic environmental monitoring, (*4*) respirator fit testing, and (*5*) respirator maintenance, inspection, cleaning, and storage. The standard also requires that the selection of respirators be made by a person knowledgeable about the workplace and the limitations associated with each type of respirator. OSHA has also issued guidelines for employers who choose to establish the voluntary use of respirators [29 CFR 1910.134 Appendix D].

Table 8–2 lists the NIOSH assigned protection factors (APF) for various classes of respirators. The APF is defined as the minimum anticipated protection provided by a properly functioning respirator or class of respirators to a given percentage of properly fitted and trained users. The APF values developed by NIOSH are based in part on laboratory studies and take into consideration a variety of factors including the inward leakage caused by penetration through the filter and leakage around the respirator face seal. The relative contributions of these two sources of inward leakage are critical because for many applications the predominant source of exposure to the respirator wearer results from leakage around the face seal (due to a poor fit) and not penetration directly through the filter media. In 2006, OSHA published updated APF values that supersede the NIOSH APF values [Federal Register 2006]. In general there is good agreement between the NIOSH and OSHA APF values, but management should consult the OSHA standard prior to using the values in Table 8–2 directly.

NIOSH is not aware of any data specific to respirator face seal leakage of nanoparticles. However, numerous studies have been conducted on larger particles and on gases/vapors with one total inward leakage (TIL) study that used nanoparticles. For example, work done by researchers at the U.S. Army RDECOM on a head-form showed that mask leakage (i.e., simulated respirator fit factor) measured using submicron aerosol challenges (0.72 μm polystyrene latex spheres) was representative of vapor challenges such as sulfur hexafluoride (SF_6) and isoamyl acetate (IAA) [Gardner et al. 2004]. Other studies using particles larger than 100 nm have shown that face seal leakage can be affected by particle size, however, the impact of this is still the subject of some debate. A recently completed laboratory study to measure TIL protection factors of four NIOSH certified N95 filtering facepiece respirator models donned by human test subjects exposed to 40–1,300 nm particles found that the minimal protection factors were observed for particles between 80–200 nm [Lee 2008]. The geometric mean of the protection factors for all four models across all particle sizes tested was 21.5; but wide model-to-model variation was observed. NIOSH is conducting a laboratory study to determine whether nanoparticle face seal leakage is consistent with the leakage observed for larger particles and gases/vapors. Results will be communicated as they become available.

NIOSH certifies respirators in accordance with 42 CFR Part 84. As noted earlier, the NIOSH RSL contains a process for selecting respirators for protection against particular hazards. The two respirator classes (air purifying respirators and powered air purifying respirators) most commonly used for protection against particulates use filter media to collect/trap particles before they reach the user's breathing zone. Among the various test methods and criteria NIOSH

uses as part of the certification process, respirator filter performance testing is the one most affected by the particle size. Since respirator users are exposed to a variety of hazards in different scenarios, respirator certification filtration testing was designed to use worst-case test conditions (e.g., different particle sizes and flow rates), so that filter performance in the workplace would not be worse. The NIOSH certification test for N-designated respirators uses a polydisperse distribution of NaCl particles with a count median diameter (CMD) of 0.075 +/-0.020 μm and a geometric standard deviation (GSD) of less than 1.86 [NIOSH 2005a]. NIOSH tests R- and P-designated respirators using a polydispersal of dioctyl phthalate (DOP) particles with a CMD of 0.185 +/-0.020 μm and a GSD of less than 1.60 [NIOSH 2005b]. For the lognormal distribution of NaCl aerosols used in the N series certification test, a broad range of particle sizes (e.g., 95% of the particles lie in the range of 22–259 nm) with a MMD of about 240 nm is used to determine whether the respirator filter performance is at least 95, 99, or 99.97% efficient. Most of the particles penetrating through the filter are measured simultaneously using a forward light scattering photometer. However, as noted in a recent review, the instrumentation used in the NIOSH certification test is not capable of measuring the light scattering of all particles less than 100 nm [Eninger et al. 2008a].

Particles larger than 0.3 μm are collected most efficiently by impaction, interception, and gravitational settling, while particles smaller than 0.3 μm are collected most efficiently by diffusion or electrostatic attraction [Hinds 1999]. In the development of the test method used for respirator certification, penetration by particles with an approximate 0.3 μm diameter was considered to be the worst case because these particles were considered to be in the range of the most penetrating particle size [Stevens and Moyer 1989; TSI 2005; NIOSH 1996]. However, in practice, the most penetrating particle size range (MPPS) for a given respirator can vary based on the type of filter media employed and the condition of the respirator. For example, the most penetrating particle size for N95 air purifying respirators containing electrostatically charged filter media can range from 50–100 nm [Martin and Moyer 2000; Richardson et al. 2005] to 30–70 nm [Balazy et al. 2006; Eninger et al. 2008b]. These test results were recently confirmed by NIOSH [Rengasamy et al. 2007] in which five different models of respirators with N95 filters were challenged with 11 different monodisperse NaCl particles ranging in size from 20–400 nm. The monodisperse aerosol penetrations showed that the MPPS was in the 40-nm range for all respirator models tested. Under the aggressive laboratory test conditions employed in the study, mean penetration levels for 40-nm particles ranged from 1.4%–5.2%, which suggested that the respirators would be effective at capturing nanoparticles in the workplace. The NIOSH study also investigated whether there was a correlation between filtration performance using the existing NIOSH certification protocol for N series air purifying respirators and the filtration performance against monodisperse particles at the MPPS. A good correlation (r = 0.95) was found (e.g., respirators that performed better using the NIOSH certification test also had higher filter efficiencies against monodisperse 40-nm nanoparticles), which is not surprising given that changes in filtration performance follow a consistent trend as a function of particle size.

According to single fiber filtration theory, below the most penetrating particle size, filtration efficiency will increase as particle size decreases. This trend will continue until the particles are so small that they behave like vapor molecules. As particles approach molecular size, they may be subject to thermal rebound effects, in which particles literally bounce through a filter. As a result, particle penetration will increase. The exact size at which thermal rebound will occur is unclear. However, a study by Heim et al. [2005] found that there was no discernable deviation from classical single-fiber theory for particles as small as 2.5-nm diameter. Subsequently, a NIOSH-funded contract with the University of Minnesota [Kim et al. 2007; Pui et al. 2006] and another study [Kim et al. 2006] showed that the penetration of nanoparticles through fibrous filter media decreased down to 2.5 nm as expected by the single fiber filtration theory. Thermal rebound phenomena were observed for nanoparticles below 2 nm diameter [Kim et al. 2006]. Recent studies provide additional data on nanoparticle penetration for NIOSH certified N95 and P100 filtering face-piece respirators [Rengasamy et al. 2008a], NIOSH certified N95 and European Certified FFP1 respirators [Huang et al. 2007], and FFP3 filter media [Golanski et al. 2008] using particles greater than 4 nm.

Based on these preliminary findings, NIOSH-certified respirators should provide the expected levels of protection if properly selected and fit tested as part of a complete respiratory protection program. However, as noted elsewhere [Rengsamy et al. 2007], in the unlikely event that the workplace exposure consists of a large percentage of particles in the most penetrating particle size range, management should take this information into account during the respirator selection process, perhaps by choosing a respirator with higher levels of filtration performance (e.g., changing from an N95 to a P100, even though the APF will remain the same) as suggested by OSHA [Federal Register 2006] or by selecting a respirator with a higher APF (e.g., full face-piece respirator or powered air purifying respirator). Dust masks, commercially available at hardware/home improvement stores, are often confused with NIOSH approved N95 filtering facepiece respirators because of their similar appearance. However, dust masks are not respirators and are not approved by NIOSH for respiratory protection. One study found that penetration of 40-nm NaCl nanoparticles range from 4.3%–81.6% for the seven dust mask models studied [Rengasamy et al. 2008b]. **Dust masks should not be used in place of NIOSH-approved respirators for protection against nanoparticles**.

NIOSH is continuing to study the protection afforded by NIOSH-certified respirators against emerging hazards such as engineered nanomaterials—including workplace-protection-factor studies—to ensure they provide expected levels of protection. NIOSH is also committed to updating 42 CFR Part 84—the regulatory language that provides NIOSH the authority to certify the performance of respirators in the United States—using a modular approach to rulemaking. Recently, NIOSH proposed the use of a TIL test as part of the respirator certification process for half-mask air purifying particulate respirators, including those having elastomeric and filtering face-pieces. The test protocol used to obtain benchmark TIL data for 101 half-face piece respirator models used 40–60 nm size ambient nanoparticles [NIOSH 2007]. Once implemented as part of the NIOSH certification process, the TIL tests should result in

half-mask respirators with increased fitting performance. Future rulemaking activities may also include revisions to the filtration test to reflect changes in filtration performance resulting from use of new technologies (e.g., electret filter media). Results will be communicated as they become available.

8.3.6 Cleanup and disposal of nanomaterials

No specific guidance is currently available on cleaning up nanomaterial spills or contamination on surfaces; however, recommendations developed in the pharmaceutical industry for the handling and cleanup of pharmaceutical compounds might be applicable to worksites where engineered nanomaterials are manufactured or used [Wood 2001]. Until relevant information is available, it would be prudent to base strategies for dealing with spills and contaminated surfaces on current good practices, together with available information on exposure risks including the relative importance of different exposure routes. Standard approaches for cleaning powder spills include using HEPA-filtered vacuum cleaners, or wiping up the powder using damp cloths or wetting the powder prior to dry wiping. Liquid spills are typically cleaned by applying absorbent materials/liquid traps.

Damp cleaning methods with soaps or cleaning oils are preferred. Cleaning cloths should be properly disposed. Use of commercially available wet or electrostatic microfiber cleaning cloths may also be effective in removing particles from surfaces with minimal dispersion into the air. Drying and reusing contaminated cloths can result in re-dispersion of particles.

Energetic cleaning methods such as dry sweeping or the using of compressed air should be avoided or only used with precautions that assure that particles suspended by the cleaning action are trapped by HEPA filters. If vacuum cleaning is employed, care should be taken that HEPA filters are installed properly and bags and filters changed according to manufacturer's recommendations.

While vacuum cleaning may prove to be effective for many applications, the following issues should be considered. Forces of attraction may make it difficult to entrain particles off surfaces with a vacuum cleaner. The electrostatic charge on particles will cause them to be attracted to oppositely charged surfaces and repelled by similarly charged surfaces. A similarly charged vacuum brush or tool may repel particles, making it difficult to capture the aerosol or even causing it to be further dispersed. Vigorous scrubbing with a vacuum brush or tool or even the friction from high flow rates of material or air on the vacuum hose can generate a charge. The vacuum cleaners recommended for cleaning copier and printer toners have

electrostatic-charge-neutralization features to address these issues.

When developing procedures for cleaning up nanomaterial spills or contaminated surfaces, consideration should be given to the potential for exposure during cleanup. Inhalation exposure and dermal exposure will likely present the greatest risks. Consideration will therefore need to be given to appropriate levels of personal protective equipment. Inhalation exposure in particular will be influenced by the likelihood of material reaerosolization. In this context, it is likely that a hierarchy of potential exposures will exist, with dusts presenting a greater inhalation exposure potential than liquids, and liquids in turn presenting a greater potential risk than encapsulated or immobilized nanomaterials and structures.

As in the case of any material spill or cleaning of contaminated surfaces, the handling and disposal of the waste material should follow existing federal, state, or local regulations.

Table 8–2. Air-purifying particulate respirators

Respirator type	NIOSH assigned protection factor	Advantages	Disadvantages
Filtering facepiece (disposable)	10	– Lightweight – No maintenance or cleaning needed – No effect on mobility	– Provides no eye protection – Can add to heat burden – Inward leakage at gaps in face seal – Some do not have adjustable head straps – Difficult for a user to do a seal check – Level of protection varies greatly among models – Communication may be difficult

(continued)

Table 8–2 (Continued). Air-purifying particulate respirators

Respirator type	NIOSH assigned protection factor	Advantages	Disadvantages
Filtering facepiece (disposable) (continued)			– Fit testing required to select proper facepiece size – Some eyewear may interfere with the fit
Elastomeric half-facepiece	10	– Low maintenance – Reusable facepiece and replaceable filters and cartridges – No effect on mobility	– Provides no eye protection – Can add to heat burden – Inward leakage at gaps in face seal – Communication may be difficult – Fit testing required to select proper facepiece size – Some eyewear may interfere with the fit
Powered with loose-fitting facepiece	25	– Provides eye protection – Offers protection for people with beards, missing dentures or facial scars – Low breathing resistance – Flowing air creates cooling effect – Face seal leakage is generally outward – Fit testing is not required – Prescription glasses can be worn – Communication easier than with elastomeric half-facepiece or full-facepiece respirators – Reusable components and replaceable filters	– Added weight of battery and blower – Awkward for some tasks – Battery requires charging – Air flow must be tested with flow device before use

(continued)

Table 8–2 (Continued). Air-purifying particulate respirators

Respirator type	NIOSH assigned protection factor	Advantages	Disadvantages
Elastomeric full-facepiece with N-100, R-100, or P-100 filters	50	– Provides eye protection – Low maintenance – Reusable facepiece and replaceable filters and cartridges – No effect on mobility – More effective face seal than that of filtering facepiece or elastomeric half-facepiece respirators	– Can add to heat burden – Diminished field-of-vision compared to half-facepiece –Inward leakage at gaps in face seal –Fit testing required to select proper facepiece size –Facepiece lens can fog without nose cup or lens treatment –Spectacle kit needed for people who wear corrective glasses
Powered with tight-fitting half-facepiece or full-facepiece	50	–Provides eye protection with full-facepiece –Low breathing resistance –Face seal leakage is generally outward –Flowing air creates cooling effect –Reusable components and replaceable filters	–Added weight of battery and blower –Awkward for some tasks –No eye protection with half-facepiece –Fit testing required to select proper facepiece size –Battery requires charging –Communication may be difficult –Spectacle kit needed for people who wear corrective glasses with full face-piece respirators –Air flow must be tested with flow device before use

Occupational health surveillance is an essential component of an effective occupational safety and health program. The unique physical and chemical properties of nanomaterials, the increasing growth of nanotechnology in the workplace, and information suggesting that exposure to some engineered nanomaterials can cause adverse health effects in laboratory animals all support consideration of an occupational health surveillance program for workers potentially exposed to engineered nanomaterials [Schulte et al. 2008a]. Continued evaluation of toxicologic research and workers potentially exposed to engineered nanomaterials is needed to inform NIOSH and other groups regarding the appropriate components of occupational health surveillance for nanotechnology workers.

NIOSH has developed interim guidance relevant to medical screening (one component of an occupational health surveillance program) for nanotechnology workers (see NIOSH Current Intelligence Bulletin: *Interim Guidance on Medical Screening and Hazard Surveillance for Workers Potentially Exposed to Engineered Nanoparticles*, www.cdc.gov/niosh/review/public/115/). Medical screening is only part of a complete safety and health management program that follows the hierarchy of controls and involves various occupational health surveillance measures. Since specific medical screening of workers exposed to engineered nanoparticles has not been extensively discussed in the scientific literature, this document is intended to fill the knowledge gap on an interim basis.

Increasing evidence indicates that exposure to some engineered nanoparticles can cause adverse health effects in laboratory animals, but no studies of workers exposed to the few engineered nanoparticles tested in animals have been published. The current body of evidence about the possible health risks of occupational exposure to engineered nanoparticles is quite small. Insufficient scientific and medical evidence now exists to recommend the specific medical screening of workers potentially exposed to engineered nanoparticles. Nonetheless, the lack of evidence on which to recommend specific medical screening does not preclude its consideration by employers interested in taking precautions beyond standard industrial hygiene measures [Schulte et al. 2008b]. If medical screening recommendations exist for chemical or bulk materials of which nanoparticles are composed, they would apply to nanoparticles as well.

Ongoing research on the hazards of engineered nanoparticles is needed along with the continual reassessment of available data to determine whether specific medical screening is warranted for workers who are producing or using nanoparticles. In the meantime, the following recommendations are provided for the management of workplaces where employees may be exposed to engineered nanoparticles in the course of their work:

- Take prudent measure to control workers' exposures to nanoparticles.

- Conduct hazard surveillance as the basis for implementing controls.

- Continue use of established medical surveillance approaches.

NIOSH will continue to examine new research findings and update its recommendations about medical screening programs for workers exposed to nanoparticles. Additionally, NIOSH is seeking comments on the strengths and weaknesses of exposure registries for workers potentially exposed to engineered nanoparticles.

NIOSH has developed a strategic plan for research on several occupational safety and health aspects of nanotechnology. The plan is available at www.cdc.gov/niosh/topics/nanotech/strat_plan.html. NIOSH has focused its research efforts in the following 10 critical topic areas to guide in addressing knowledge gaps, developing strategies, and providing recommendations.

1. **Exposure Assessment**

    — Determine key factors that influence the production, dispersion, accumulation, and re-entry of nanomaterials into the workplace.

    — Determine how possible exposures to nanomaterials differ by work process.

    — Assess possible exposure when nanomaterials are inhaled or settle on the skin.

2. **Toxicity and Internal Dose**

    — Investigate and determine the physical and chemical properties (e.g., size, shape, solubility, surface area, oxidant generation potential, surface functionalization, surface charge, chemical composition) that influence the potential toxicity of nanomaterials.

    — Determine the deposition pattern of nanoparticles in the lung and their translocation to the interstitium and to extrapulmonary organs.

    — Evaluate short- and long-term effects of pulmonary exposure to nanomaterials in various organ systems and tissues (e.g., lungs, brain, cardiovascular).

    — Determine if intratracheal instillation or pharyngeal aspiration can mimic the biological response to inhalation exposure to nanomaterials.

    — Determine the dermal effects of topical exposure to nano-objects, whether these nano-objects can penetrate into the skin, and whether they can cause immune alterations.

    — Determine the genotoxic and carcinogenic potential of nano-objects.

    — Determine biological mechanisms for potential toxic effects.

    — Determine whether in vitro screening tests can be predictive on in vivo response.

    — Create and integrate models to help assess potential hazards.

    — Determine whether a measure other than mass is more appropriate for determining toxicity.

3. **Epidemiology and Surveillance**

    — Evaluate existing exposure and health data for workers employed in workplaces where nanomaterials are produced and used, with emphasis on improving our understanding of the value and utility of establishing exposure registries for workers potentially exposed to engineered nanomaterials.

— Assess the feasibility of industry-wide exposure and epidemiological studies of workers exposed to engineered nanomaterials, with emphasis on workers potentially exposed to engineered carbonaceous nanomaterials.

— Integrate nanotechnology safety and health issues into existing hazard surveillance mechanisms and continue reassessing guidance related to occupational health surveillance for workers potentially exposed to engineered nanomaterials.

— Build on existing public health geographical information systems and infrastructure to enable effective and economic development of methods for sharing nanotechnology safety and health data.

4. Risk Assessment

— Determine how existing exposure-response data for fine and ultrafine particles (human or animal) may be used to identify the potential hazards and estimate the potential risks of occupational exposure to nanomaterials.

— Develop a framework for assessing the potential hazards and risks of occupational exposure to nanomaterials, using new toxicologic data on engineered nanomaterials and standard risk assessment models and methods.

5. Measurement Methods

— Evaluate methods used to measure the mass of respirable particles in the air and determine whether this measurement can be used to measure nanomaterials.

— Develop and field-test practical methods to accurately measure airborne nanomaterials in the workplace.

— Develop, test, and evaluate systems to compare and validate sampling.

6. Engineering Controls and Personal Protective Equipment

— Evaluate the effectiveness of engineering controls in reducing occupational exposures to nano-aerosols and developing new controls when needed.

— Evaluate the suitability of control-banding techniques when additional information is needed and evaluate the effectiveness of alternative materials.

— Evaluate and improve current personal protective equipment.

— Develop recommendations (e.g., use of respiratory protection) to prevent or limit occupational exposures to nanomaterials.

7. Fire and Explosion Safety

— Identify physical and chemical properties that contribute to dustiness, combustibility, flammability, and conductivity of nanomaterials.

— Recommend alternative work practices to eliminate or reduce work place exposures to nanomaterials.

8. Recommendations and Guidance

— Use the best available science to make interim recommendations for workplace safety and health practices during the production, use, and handling of nanomaterials.

— Evaluate and update mass-based occupational exposure limits for airborne particles to ensure good, continuing precautionary practices.

9. Communication and Information

— Establish partnerships to allow for identification and sharing of research needs, approaches, and results.

— Develop and disseminate training and education materials to workers, employers, and occupational safety and health professionals.

10. Applications

— Identify uses of nanotechnology for application in occupational safety and health.

— Evaluate and disseminate effective applications to workers, employers, and occupational safety and health professionals.

ACGIH [2001]. Industrial ventilation: a manual of recommended practice. Cincinnati, OH: American Conference of Governmental Industrial Hygienists.

Adams RJ, Bray D [1983]. Rapid transport of foreign particles microinjected into crab axons. Nature 303:718–720.

Ambroise D, Wild P, Moulin J-J [2007]. Update of a meta-analysis on lung cancer and welding. Scand J Work Environ Health 32(1):22–31.

Antonini JM [2003]. Health effects of welding. Crit Rev Toxicol 33(1):61–103.

ASTM Subcommittee F23.40 [2003]. Standard test method for resistance of materials used in protective clothing to penetration by blood-borne pathogens using Phi–X174 bacteriophage penetration as a test system. West Conshohocken, PA: American Society for Testing and Materials, ASTM F1671–03.

ASTM Committee E2535–07 [2007]. Standard guide for handling unbound engineered nanoscale particles in occupational settings. West Conshohocken, PA: ASTM International.

Balazy A, Toivola M, Reponen T, Podgorski A, Zimmer A, Grinshpun SA [2006]. Manikin-based filtration performance evaluation of filtering-facepiece respirators challenged with nanoparticles. Ann Occup Hyg 50(3):259–269.

Baltensperger U, Gaggeler HW, Jost DT [1988]. The epiphaniometer, a new device for continuous aerosol monitoring. J Aerosol Sci 19(7):931–934.

Barlow PG, Clouter-Baker AC, Donaldson K, MacCallum J, Stone V [2005]. Carbon black nanoparticles induce type II epithelial cells to release chemotaxins for alveolar macrophages. Part Fiber Toxicol 2(14).

Baron PA, Deye GJ, Chen B, Schwegler-Berry D, Shvedova AA, Castranova V [in press]. Aerosolization of single-walled carbon nanotubes for an inhalation study. Inhal Toxicol.

Bergman W, Garr J, Fearon D [1989]. Aerosol penetration measurements through protective clothing in small scale simulation tests. Lawrence Livermore National Laboratory. Presented at: 3rd International Symposium on Protection against Chemical Warfare Agents, UMEA, Sweden, June 11–16.

Borm PJA, Schins RPF, Albrecht C [2004]. Inhaled particles and lung cancer, Part B: Paradigms and risk assessment. Int J Cancer 110:3–14.

Bowler RM, Roels HA, Nakagawa S, Drezgic M, Diamond E, Park R, Koller W, Bowler RP, Mergler D, Bouchard M, Smith D, Gwiazda R, Doty RL [2007]. Dose-effect relationships between manganese exposure and neurological, neuropsychological and pulmonary function in confined space bridge welders. Occup Environ Med 64(3):167–177.

British Standards Institute [2008]. Nanotechnologies, Part 2: Guide to safe handling and disposal of manufactured nanomaterials. PD 6699–2:2007 [www.bsi-global.com/upload/Standards%20&%20 Publications/Nanotechnologies/PD6699-2.pdf].

Brouwer DH, Gijsbers JHJ, Lurvink MWM [2004]. Personal exposure to ultrafine particles in the workplace: exploring sampling techniques and strategies. Ann Occup Hyg 48(5):439–453.

Brown DM, Wilson MR, MacNee W, Stone V, Donaldson K [2001]. Size-dependent proinflammatory effects of ultrafine polystyrene particles: a role for surface area and oxidative stress in the enhanced activity of ultrafines. Toxicol App Pharmacol 175(3):191–199.

Brown JS, Zeman KL, Bennett WD [2002]. Ultrafine particle deposition and clearance in the healthy and obstructed lung. Am J Respir Crit Care Med 166:1240–1247.

Brunauer S, Emmett PH, Teller E [1938]. Adsorption of gases in multimolecular layers. J Am Chem Soc 60:309.

Burton J [1997]. General methods for the control of airborne hazards. In: DiNardi SR, ed. The occupational environment—its evaluation and control. Fairfax, VA: American Industrial Hygiene Association.

References

Castranova V [2000]. From coal mine dust to quartz: mechanisms of pulmonary pathogenicity. Inhal Tox 12(Suppl 3):7–14.

Daigle CC, Chalupa DC, Gibb FRMorrow PE, Oberdorster G, Utell MJ, Frampton MW [2003]. Ultrafine particle deposition in humans during rest and exercise. Inhal Toxicol 15(6):539–552.

De Lorenzo AJD [1970]. The olfactory neuron and the blood-brain barrier. In: Wolstenholme GEW, Knight J, eds. Taste and smell in vertebrates. CIBA Foundation Symposium series. London: J&A Churchill, pp. 151–176.

Dockery DW, Pope CA, Xu X, Spengler JD, Ware JH, Fay ME, Ferris BG, Speizer BE [1993]. An association between air pollution and mortality in six U.S. cities. N Engl J Med 329(24):1753–1759.

Donaldson K, Li XY, MacNee W [1998]. Ultrafine (nanometer) particle mediated lung injury. J Aerosol Sci 29(5–6):553–560.

Donaldson K, Tran CL [2002]. Inflammation caused by particles and fibers. Inhal Toxicol 14(1):5–27.

Donaldson K, Stone V [2003]. Current hypotheses on the mechanisms of toxicity of ultrafine particles. Ann 1st Super Sanita 39(3):405–410.

Donaldson K, Tran L, Jimenez LA, Duffin R, Newby DE, Mills N, MacNee W, Stone V [2005]. Combustion-derived nanoparticles: a review of their toxicology following inhalation exposure. Part Fibre Toxicol. 2(10):14.

Donaldson K, Aitken R, Tran L, Stone V, Duffin R, Forrest G, Alexander A [2006]. Carbon nanotubes: a review of their properties in relation to pulmonary toxicology and workplace safety. Toxicol Sci 92(1):5–22.

Driscoll KE [1996]. Role of inflammation in the development of rat lung tumors in response to chronic particle exposure. In: Mauderly JL, McCunney RJ, eds. Particle overload in the rat lung and lung cancer: implications for human risk assessment. Philadelphia, PA: Taylor & Francis, pp. 139–152.

Duffin R, Tran CL, Clouter A, Brown DM, MacNee W, Stone V, Donaldson K [2002]. The importance of surface area and specific reactivity in the acute pulmonary inflammatory response to particles. Ann Occup Hyg 46:242–245.

Duffin R, Tran L, Brown D, Stone V, Donaldson K [2007]. Proinflammogenic effects of low-toxicity and metal nanoparticles in vivo and in vitro: highlighting the role of particle surface area and surface reactivity. Inhal Toxicol 19(10):849–856.

Edlich RF, Suber F, Neal JG, Jackson EM, Williams FM [1999]. Integrity of powder-free gloves to bacteriophage penetration. J Biomed Mater Res B Appl Biomater 48:755–758.

Elder A, Gelein R, Silva V, Feikert T, Opanashuk L, Carter J, Potter R, Maynard A, Ito Y, Finkelstein J, Oberdörster G [2006]. Translocation of inhaled ultrafine manganese oxide particles to the central nervous system. Environ Health Perspect 114:1172–1178.

Eninger RM, Honda T, Reponen T, McKay R, Grinshpun SA [2008a]. What does respirator certification tell us about filtration of ultrafine particles? J Occup Environ Hyg 5:286–295.

Eninger RM, Honda T, Adhikari A, Heinonen-Tanski H, Reponen T, Grinshpun SA [2008b]. Filter performance of N99 and N95 facepiece respirators against viruses and ultrafine particles. Ann Occup Hyg 52(5):385–396.

71 Fed. Reg. 50121–50192 [2006]. Occupational Safety and Health Administration: assigned protection factors, final rule [www.osha.gov/pls/oshaweb/owadisp.show_document?p_table=FEDERAL_REGISTER&p_id=18846].

European Committee for Standardization [2002]. CEN 943–1:2002: Protective clothing against liquid and gaseous chemicals, aerosols and solid particles. Performance requirements for ventilated and non-ventilated "gas-tight" (Type 1) and "non-gas-tight" (Type 2) chemical protective suits. Brussels: European Committee for Standardization.

Evans DE, Heitbrink WA, Slavin TJ, Peters TM [2008]. Ultrafine and respirable particles in an automotive grey iron foundry. Ann Occup Hyg 52(1):9–21.

Frampton MW, Stewart JC, Oberdorster G, Morrow PE, Chalupa D, Pietropaoli AP, Frasier LM, Speers DM, Cox C, Huang LS, Utell MJ [2006]. Inhalation of ultrafine particles alters blood leukocyte expression of adhesion molecules in humans. Environ Health Perspect 114(1):51–58.

Fuchs NA [1964]. The mechanics of aerosols. Oxford, England: Pergamon Press.

Gao P, King WP, Shaffer R [2007]. Review of chamber design requirements for testing of personal protective clothing ensembles. J Occup & Environ Hyg 4(8):562–571.

Gardiner K, van Tongeren M, Harrington M [2001]. Respiratory health effects from exposure to carbon black: results of the phase 2 and 3 cross sectional studies in the European carbon black manufacturing industry. Occup Environ Med 58(8):496–503.

Gardner P, Hofacre K, Richardson A [2004]. Comparison of simulated respirator fit factors using aerosol and vapor challenges. J Occup and Environ Hyg 1(1):29–38.

Garshick E, Laden F, Hart JE, Rosner B, Smith TJ, Dockery DW, Speizer FE [2004]. Lung cancer in railroad workers exposed to diesel exhaust. Environ Health Perspect 112(15):1539–1543.

Garshick E, Laden F, Hart JE, Smith TJ, Rosner B [2006]. Smoking imputation and lung cancer in railroad workers exposed to diesel exhaust. Am J Ind Med 49(9):709–718.

Geiser M, Rothen-Rutishauser B, Kapp N, Schurch S, Kreyling W, Schulz H, Semmler M, Im Hof V, Heyder J, Gehr P [2005]. Ultrafine particles cross cellular membranes by nonphagocytic mechanisms in lungs and in cultured cells. Environ Health Perspect 113(11):1555–1560.

Golanski L, Guiot A, Tardif F [2008]. Are conventional protective devices such as fibrous filter media, respirator cartridges, protective clothing and gloves also efficient for nanoaerosols? Brussels, Germany: European Community, Nanotech2 Project.

Goldstein M, Weiss H, Wade K, Penek J, Andrews L, Brandt-Rauf P [1987]. An outbreak of fume fever in an electronics instrument testing laboratory. J Occup Med 29:746–749.

Granier JJ, Pantoya ML [2004]. Laser ignition of nanocomposite thermites. Combust Flame 138:373–382.

Han JH, Lee EJ, Lee JH, So KP, Lee YH, Bae GN, Lee S-B, Ji JH, Cho MH, Yu IJ [2008]. Monitoring multiwalled carbon nanotube exposure in carbon nanotube research facility. Inhal Toxicol 20(8):741–749.

Hart JE, Laden F, Schenker MB, Garshick E [2006]. Chronic obstructive pulmonary disease mortality in diesel-exposed railroad workers. Environ Health Perspect 114(7):1013–1017.

HEI [2000]. Reanalysis of the Harvard Six Cities Study and the American Cancer Society Study of Particulate Air Pollution and Mortality. Cambridge, MA: Health Effects Institute.

Heim M, Mullins B, Wild M, Meyer J, Kasper, G [2005]. Filtration efficiency of aerosol particles below 20 nanometers. Aerosol Sci Tech 39:782–789.

Heinrich U, Fuhst R, Rittinghausen S, Creutzenberg O, Bellmann B, Koch W, Levsen K [1995]. Chronic inhalation exposure of wistar rats and 2 different strains of mice to diesel-engine exhaust, carbon-black, and titanium-dioxide. Inhal Toxicol 7(4):533–556.

Heitbrink WA, Evans DE, Peters TM, Slavin TJ [2007]. The characteristion and mapping of very fine particles in an engine machining and assembly facility. J Occup Environ Hyg 4:341–351.

Heitbrink WA, Evans DE, Ku BK, Maynard AD, Slavin TJ, Peters TM [forthcoming]. Relationships between particle number, surface area, and respirable mass concentration in automotive engine manufacturing. J Occup Environ Hyg.

Hinds WC [1999]. Aerosol technology: properties, behavior, and measurement of airborne particles. 2nd ed. New York: Wiley-Interscience.

Hofacre KC [2006]. Aerosol penetration of fabric swatches. Unpublished paper presented at the Elevated Wind Aerosol Conference, Arlington, VA, September 2006.

Hoshino A, Fujioka K, Oku T, Suga M, Ssaki Y, Ohta T [2004]. Physicochemical properties and cellular toxicity of nanocrystal quantum dots depend on their surface modification. Nano Lett 4(11):2163–2169.

HSE [2004]. Horizon scannon information sheet on nanotechnology. Sudbury, Suffolk, United Kingdom: Health and Safety Executive [www.hse.gov/pubns/hsin1.pdf].

Huang SH, Chen CW, Chang CP, Lai CY, Chen CC [2007]. Penetration of 4.5 nm to 10 μm aerosol particles through fibrous filters. J Aero Sci 38:719–727.

Hunter DD, Dey RD [1998]. Identification and neuropeptide content of trigeminal neurons innervating the rat nasal epithelium. Neuroscience 83(2):591–599.

References

Ibald-Mulli A, Wichmann HE, Kreyling W, Peters A [2002]. Epidemiological evidence on health effects of ultrafine particles. J Aerosol Med Depos 15(2):189–201.

International Council on Nanotechnology (ICON) [2006]. A survey of current practices in the nanotechnology workplace. [http://cohesion.rice.edu/CentersAndInst/ICON/emplibrary/ICONNanotechSurveyFullReduced.pdf].

ICRP [1994]. Human respiratory tract model for radiological protection. Oxford, England: Pergamon, Elsevier Science Ltd., International Commission on Radiological Protection, Publication No. 66.

ISO [2004a]. Protective clothing for use against solid particulates. Part 2: Test method of determination of inward leakage of aerosols of fine particles into suits. ISO Standard 13982 2. Geneva, Switzerland: International Organization for Standardization.

ISO [2004b]. ISO 16604:2004: Clothing for protection against contact with blood and body fluids: determination of resistance of protective clothing materials to penetration by blood-borne pathogens; test method using Phi-X 174 bacteriophage. Geneva, Switzerland: International Organization for Standardization.

ISO [1995]. Air quality; particle size fraction definitions for health-related sampling, ISO 7708. Geneva, Switzerland: International Organization for Standardization.

ISO [2006]. Workplace atmospheres; ultrafine, nanoparticle and nano-structured aerosols; exposure characterization and assessment. Geneva, Switzerland: International Standards Organization. Document no. ISO/TC 146/SC 2/WG1 N324, 32 pages.

Jaques PA, Kim CS [2000]. Measurement of total lung deposition of inhaled ultrafine particles in healthy men and women. Inhal Toxicol 12(8):715–731.

Jiang J, Oberdorster G, Elder A, Gelein R, Mercer P, Biswas P [2008]. Does nanoparticle activity depend upon size and crystal phase? Nanotox 2(1):33–42.

Johnston CJ, Finkelstein JN, Mercer P, Corson N, Gelein R, Oberdorster G [2000]. Pulmonary effects induced by ultrafine PTFE particles. Toxicol Appl Pharmacol 168:208–215.

Keskinen J, Pietarinen K, Lehtimaki M [1992]. Electrical low pressure impactor. J Aerosol Sci 23:353–360.

Kim CS, Jaques PA [2004]. Analysis of total respiratory deposition of inhaled ultrafine particles in adult subjects at various breathing patterns. Aerosol Sci Technol 38:525–540.

Kim CS, Bao L, Okuyama K, Shimada S, Niinuma H [2006]. Filtration efficiency of a fibrous filter for nanoparticles. J Nanopart Res 8:215 –221.

Kim SE, Harrington MS, Pui DYH [2007]. Experimental study of nanoparticles penetration through commercial filter media. J Nanopart Res 9:117–125.

Kreiss K, Mroz MM, Zhen B, Wiedemann H, Barna B [1997]. Risks of beryllium disease related to work processes at a metal, alloy, and oxide production plant. Occup Environ Med 54(8):605–612.

Kreyling WG, Semmler M, Erbe F, Mayer P, Takenaka S, Schulz H, Oberdorster G, Ziesenis A [2002]. Translocation of ultrafine insoluble iridium particles from lung epithelium to extrapulmonary organs is size dependent but very low. J Toxicol Environ Health 65(20):1513–1530.

Kreyling WG, Semmler-Behnke M, Moller W [2006]. Ultrafine particle-lung interactions: does size matter? J Aerosol Med 19(1):74–83.

Ku BK, Maynard AD [2005]. Comparing aerosol surface-area measurements of monodisperse ultrafine silver agglomerates by mobility analysis, transmission electron microscopy and diffusion charging. J Aerosol Sci 36:1108–1124.

Ku BK, Maynard AD [2006]. Generation and investigation of airborne silver nanoparticles with specific size and morphology by homogeneous nucleation, coagulation and sintering. J Aerosol Sci 37:452–470.

Kuhlbusch TAJ, Neumann S, Fissan H [2004]. Number size distribution, mass concentration, and particle composition of PM1, PM2.5, and PM10 in bag filling areas of carbon black production. J Occup Environ Hyg 1(10):660–671.

Lam CW, James JT, McCluskey R, Hunter RL [2004]. Pulmonary toxicity of single-wall carbon nanotubes in mice 7 and 90 days after intratracheal instillation. Toxicol Sci 77:126–134.

Lam CW, James JT, McCluskey R, Arepalli S, Hunter RL [2006]. A review of carbon nanotube toxicity and assessment of potential occupational and environmental health risks. Crit Rev Toxicol 36:189–217.

Lee KP, Trochimowicz HJ, Reinhardt CF [1985]. Pulmonary response of rats exposed to titanium dioxide (TiO$_2$) by inhalation for two years. Toxicol Appl Pharmacol 79:179–192.

Lee CH, Guo YL, Tsai PJ, Chang HY, Chen CR, Chen CW, Hsiue TR [1997]. Fatal acute pulmonary oedema after inhalation of fumes from polytetrafluoroethylene (PTFE). Eur Res J 10:1408–1411.

Lee SA, Grinshpun SA, Reponen T [2008]. Respiratory performance offered by N95 respirators and surgical masks: Human subject evaluation with NaCl aerosol representing bacterial and viral particle size range. Ann Occup Hyg 52(3):177–85. Epub 2008 Mar 7.

Lippmann M [1977]. Regional deposition of particles in the human respiratory tract. In: Lee DHK, Murphy S, eds. Handbook of physiology; section IV, environmental physiology. Philadelphia, PA: Williams and Wilkins, pp. 213–232.

Li N, Sioutas C, Cho A, Schmitz D, Misra C, Sempf J, Wang MY, Oberley T, Froines J, Nel A [2003]. Ultrafine particulate pollutants induce oxidative stress and mitochondrial damage. Environ Health Perspect 111(4):455–460.

Li Z, Hulderman T, Salmen R, Chapman R, Leonard SS, Young S-H, Shvedova A, Luster MI, Simeonova PP [2007]. Cardiovascular effects of pulmonary exposure to single-wall carbon nanotubes. Environ Health Perspect 115:377–382.

Lison D, Lardot C, Huaux F, Zanetti G, Fubini B [1997]. Influence of particle surface area on the toxicity of insoluble manganese dioxide dusts. Arch Toxicol 71(12):725–729.

Long TC, Saleh N, Tilton RD, Lowry GV, Veronesi B [2006]. Titanium dioxide (P25) produces reactive oxygen species in immortalized brain microglia (BV2): implications for nanoparticle neurotoxicity. Environ Sci & Technol 40(14):4346–4352. Epub 16 June 2006.

Lovric J, Bazzi HS, Cuie Y, fortin, GRA, Winnik FM, D Maysinger [2005]. Differences in subcellular distribution and toxicity of green and red emitting CdTe quantum dots. J Mol Med 83(5):377–385. Epub 2005 Feb 2.

Lux Research [2007]. The Nanotech Report, 5th Edition. New York: Lux Research.

Marple VA, Olson BA, Rubow KL [2001]. Inertial, gravitational, centrifugal, and thermal collection techniques. In: Baron PA, Willeke K, eds. Aerosol measurement: principles, techniques and applications. New York: John Wiley and Sons, pp. 229–260.

Martin S, Moyer E [2000]. Electrostatic respirator filter media: filter efficiency and most penetrating particle size effects. App Occup Environ Hyg 15(8):609–617.

Maynard AD [2003]. Estimating aerosol surface area from number and mass concentration measurements. Ann Occup Hyg 47(2):123–144.

Maynard AD, Baron PA, Foley M, Shvedova AA, Kisin ER, Castranova V [2004]. Exposure to carbon nanotube material: aerosol release during the handling of unrefined single walled carbon nanotube material. J Toxicol Environ Health 67(1):87–107.

Maynard AM, Kuempel ED [2005]. Airborne nanostructured particles and occupational health. J Nanoparticle Research 7(6):587–614.

Mercer RR, Scabilloni J, Wang L, Kisin E, Murray AR, Schwegler-Berry D, Shvedova AA, Castranova V [2008]. Alteration of depositon pattern and pulmonary response as a result of improved dispersion of aspirated single walled carbon nanotubes in a mouse model. Am J Physiol Lung Cell Mol Physiol 294:L87–L97.

Methner MM, Birch ME, Evans DE, Ku BK, Crouch KG, Hoover MD [2007]. Mazzulxeli LF, ed: Case study: Identification and characterization of potential sources of worker exposure to carbon nanofibers during polymer composite laboratory operations. J Occup Environ Hyg 4(12):D125–D130.

Methner MM [2008]. Engineering case reports: Old L, ed. Effectiveness of local exhaust ventilation (LEV) in controlling engineered nanomaterial emissions during reactor cleanout operations. J Occup Environ Hyg 5(6):D63–D69.

Misra C, Singh M, Shen S, Sioutas C, Hall PM [2002]. Development and evaluation of a personal cascade impactor sampler (PCIS). J Aerosol Sci 33(7):1027–1048.

Moller W, Hofer T, Ziesenis A, Karg E, Heyder J [2002]. Ultrafine particles cause cytoskeletal dysfunctions in macrophages. Toxicol Appl Pharmacol 182(3):197–207.

References

Moller W, Brown DM, Kreyling WG, Stone V [2005]. Ultrafine particles cause cytoskeletal dysfunctions in macrophages: role of intracellular calcium. Part Fibre Toxicol 2(7):12.

Monteiro-Riviere NA, Nemanich RJ, Inman AO, Wang YY, Riviere JE [2005]. Multi-walled carbon nanotube interactions with human epidermal keratinocytes. Toxicol Lett 155(3):377–384.

Moshammer H, Neuberger M [2003]. The active surface of suspended particles as a predictor of lung function and pulmonary symptoms in Austrian school children. Atmos Environ 37(13):1737–1744.

Mossman B, Churg A [1998]. Mechanisms in the pathogenesis of asbestosis and silicosis. Am J Respir Crit Care Med 157:1666–1680.

Muller J, Huaux F, Moreau N, Misson P, Heilier J F, Delos M, Arras M, Fonseca A, Nagy JB, Lison D [2005]. Respiratory toxicity of multi-wall carbon nanotubes. Toxicol Appl Pharmacol 207:221–231.

Murray AR, Kisin E, Kommineni C, Kagan VE, Castranova V, Shvedova AA [2007]. Single-walled carbon nanotubes induce oxidative stress and inflammation in skin. Toxicologist 96:A1406.

National Fire Protection Association (NFPA) [2007]. NFPA 1994 standard on protective ensembles for first responders to CBRN terrorism incidents.

Nel A, Xia T, Madlen L, Li W [2006]. Toxic potential of materials at the nanolevel. Science 311:622–627.

Nemmar A, Hoet PHM, Vanquickenborne B, Dinsdale D, Thomeer M, Hoylaerts MF, Vanbilloen H, Mortelmans L, Nemery B [2002]. Passage of inhaled particles into the blood circulation in humans. Circulation 105:411–414.

NIOSH [1990]. NIOSH testimony on the Occupational Safety and Health Administration proposed rule on health standards: methods of compliance. Cincinnati, OH: U.S. Department of Health and Human Services, Centers for Disease Control, National Institute for Occupational Safety and Health.

NIOSH [1994]. NIOSH manual of analytical methods (NMAM®). 4th ed. By Schlecht PC, O'Connor PF, eds. Cincinnati, OH: U.S. Department of Health and Human Services, Centers for Disease Control and Prevention, National Institute for Occupational

Safety and Health, DHHS (NIOSH) Publication 94–113 [www.cdc.gov/niosh/nmam/].

NIOSH [1996]. NIOSH guide to the selection and use of particulate respirators certified under 42 CFR 84. Cincinnati, OH: U.S. Department of Health and Human Services, Centers for Disease Control and Prevention, National Institute for Occupational Safety and Health, DHHS (NIOSH) Publication No. 96–101.

NIOSH [2003]. Filtration and air-cleaning systems to protect building environments. Cincinnati, OH: U.S. Department of Health and Human Services, Centers for Disease Control and Prevention, National Institute for Occupational Safety and Health, DHHS (NIOSH) Publication No. 2003–136.

NIOSH [2004]. NIOSH respirator selection logic. Cincinnati, OH: U.S. Department of Health and Human Services, Centers for Disease Control and Prevention, National Institute for Occupational Safety and Health, DHHS (NIOSH) Publication No. 2005–100 [www.cdc.gov/niosh/docs/2005-100/].

NIOSH [2005a]. Procedure No. RCT–APR–STP–0057, 0058, 0059, Revision 1.1. Cincinnati, OH: Department of Health and Human Services, Centers for Disease Control and Prevention, National Institute for Occupational Safety and Health [www.cdc.gov/niosh/npptl/stps/pdfs/RCT-APR-0057%2058%2059.pdf].

NIOSH [2005b]. Procedure No. RCT–APR–STP–0051, 0052, 0053, 0054, 0055, 0056, Revision 1.1. Cincinnati, OH: Department of Health and Human Services, Centers for Disease Control and Prevention, National Institute for Occupational Safety and Health [www.cdc.gov/niosh/npptl/stps/pdfs/RCT-APR-0051%2052%2053%2054%2055%2056.pdf].

NIOSH [2009]. Current intelligence bulletin: interim guidance on medical screening of workers potentially exposed to engineered nanoparticles. Cincinnati, OH: U.S. Department of Health and Human Services, Centers for Disease Control and Prevention, National Institute for Occupational Safety and Health [www.cdc.gov/niosh/docs/2009-116/].

NIOSH [2007]. Docket Number 036: Total Inward Leakage Test for Half-Mask Air-Purifying Particulate Respirators; Total Inward Leakage Test for Half-mask Air-purifying Particulate Respirators. Procedure No.

RCT–APR–STP–0068, January 31, 2007 [www.cdc.gov/niosh/docket/NIOSHdocket0036.html].

Oberdörster G, Yu [1990]. The carcinogenic potential of inhaled diesel exhaust: a particle effect? J Aerosol Sci *21*:S397–S401.

Oberdörster G, Ferin J, Gelein R, Soderholm SC, Finkelstein J [1992]. Role of the alveolar macrophage in lung injury—studies with ultrafine particles. Environ Health Perspect *97*:193–199.

Oberdörster G, Ferin J, Lehnert BE [1994a]. Correlation between particle-size, in-vivo particle persistence, and lung injury. Environ Health Perspect *102*(S5):173–179.

Oberdörster G, Ferin J, Soderholm S, Gelein R, Cox C, Baggs R, Morrow PE [1994b]. Increased pulmonary toxicity of inhaled ultrafine particles: due to lung overload alone? Ann Occup Hyg *38*(Suppl 1):295–302.

Oberdörster G, Gelein RM, Ferin J, Weiss B [1995]. Association of particulate air pollution and acute mortality: involvement of ultrafine particles? Inhal Toxicol *7*(1):111–124.

Oberdörster G, Sharp Z, Atudorei V, Elder A, Gelein R, Lunts A, Kreyling W, Cox C [2002]. Extrapulmonary translocation of ultrafine carbon particles following whole-body inhalation exposure of rats. J Toxicol Environ Health *65* Part A(20):1531–1543.

Oberdörster G, Sharp Z, Atudorei V, Elder A, Gelein R, Kreyling W, Cox C [2004]. Translocation of inhaled ultrafine particles to the brain. Inhal Toxicol *16*(6–7):437–445.

Oberdörster G, Oberdörster E, Oberdörster J [2005a]. Nanotoxicology: an emerging discipline evolving from studies of ultrafine particles. Environ Health Perspect *113*(7):823–839.

Oberdörster G, Maynard A, Donaldson K, Castranova V, Fitzpatrick J, Ausman K, Carter J, Karn B, Kreyling W, Lai D, Olin S, Monteiro-Riviere N, Warheit D, Yang H [2005b]. Principles for characterizing the potential human health effects from exposure to nanomaterials: elements of a screening strategy. Part Fibre Toxicol *2*:8.

Park RM, Bowler RM, Eggerth DE, Diamond E, Spencer KJ, Smith D, Gwiazda R [2006]. Issues in neurological risk assessment for occupational exposures: the Bay Bridge welders. Neurotoxicology *27*(3):373–384.

Penttinen P, Timonen KL, Tiittanen P, Mirme A, Russkanen J, Pekkanen J [2001]. Ultrafine particles in urban air and respiratory health among adult asthmatics. Eur Respir J *17*(3):428–435.

Peters A, Dockery DW, Heinrich J, Wichmann HE [1997]. Short-term effects of particulate air pollution on respiratory morbidity in asthmatic children. Eur Respir J *10*(4):872–879.

Peters A, von Klot S, Heier M, Trentinaglia I, Hormann A, Wichmann HE, Lowel H [2004]. Exposure to traffic and the onset of myocardial infarction. N Engl J Med *351*(17):1721–1730.

Peters TM, Heitbrink WA, Evans DE, Slavin TJ, Maynard AD [2006]. The mapping of fine and ultrafine particle concentrations in an engine machining and assembly facility. Ann Occup Hyg *50*(3):249–257.

Plog BA, Niland J, Quinlan PJ, eds. [2002]. Fundamentals of industrial hygiene, 5th ed. Itasca, IL: National Safety Council.

Poland CA, Duffin R, Kinloch I, Maynard A, Wallace WAH, Seaton A, Stone V, Brown S, MacNee W, Donaldson K [2008]. Carbon nanotubes introduced into the abdominal cavity of mice show asbestos-like pathogenicity in a pilot study. Nat Nanotech *3*(7):423–428. Epub 2008 May 20. [www.nature.com/naturenanotechnology].

Pope CA, Burnett RT, Thun MJ, Calle EE, Krewski E, Ito K, Thurston GD [2002]. Lung cancer, cardiopulmonary mortality and long term exposure to fine particulate air pollution. JAMA *287*(9):1132–1141.

Pope CA, Burnett RT, Thurston GD, Thun MJ, Calle EE, Krewski D, Godleski JJ [2004]. Cardiovascular mortality and long-term exposure to particulate air pollution: epidemiological evidence of general pathophysiological pathways of disease. Circulation *109*(1):71–74.

Porter AE, Gass M, Muller K, Skepper JN, Midgley P, Welland M [2007a]. Direct imaging of single-walled carbon nanotubes in cells. Nat Nanotech *2*:713–717.

Porter AE, Gass M, Muller K, Skepper JN, Midgley P, Welland M [2007b]. Visualizing the uptake of C60 to the cytoplasm and nucleus of human monocyte-derived macrophage cells using energy-filtered transmission electron microscopy and electron tomography. Environ Sci Technol 41(8):3012–3017.

Pott F, Roller M [2005]. Carcinogenicity study with nineteen granular dusts in rats. Eur J Oncol 10(4):249–281.

Pritchard DK [2004]. Literature review—explosion hazards associated with nanopowders. United Kingdom: Health and Safety Laboratory, HSL/2004/12.

Pui DYH, Kim SC [2006]. Penetration of nanoparticles through respirator filter media. Minneapolis, MN: University of Minnesota, Mechanical Engineering Department, Particle Technology Laboratory. NIOSH Contract No. 254–2005–M–11698 for National Personal Protective Technology Division.

Ratherman S [1996]. Methods of control. In: Plog D, ed. Fundamentals of industrial hygiene. Itasca, IL: National Safety Council.

Rengasamy S, Verbofsky R, King WP, Shaffer RE [2007]. Nanoparticle penetration through NIOSH-approved N95 filtering-facepiece respirators. J Int Soc Res Prot 24:49–59.

Rengasamy S, King WP, Eimer B, Shaffer RE [2008a]. Filtration performance of NIOSH-approved N95 and P100 filtering-facepiece respirators against 4–30 nanometer size nanoparticles. J Occup Environ Hyg 5(9):556–564.

Rengasamy S, Eimer BC, Shaffer RE [2008b]. Nanoparticle filtration performance of commercially available dust masks. J Int Soc Res Prot 25:27.

Renwick LC, Brown D, Clouter A, Donaldson K [2004]. Increased inflammation and altered macrophage chemotactic responses caused by two ultrafine particles. Occup Environ Med 61:442–447.

Richardson, AW, Eshbaugh, JP, Hofacre, KC, Gardner PD [2005]. Respirator filter efficiency testing against particulate and biological aerosols under moderate to high flow rates. Columbus, OH: Battelle Memorial Institue. Contract Report No. SP0700–00–D–3180.

Ruckerl R, Ibald-Mulli A, Koenig W, Schneider A, Woelke G, Cyrys J, Heinrich J, Marder V, Frampton M, Wichmann HE, Peters A [2006]. Air pollution and markers of inflammation and coagulation in patients with coronary heart disease. Am J Respir Crit Care Med 173(4):432–441.

Ryman-Rasmussen JP, JE Riviere JE, NA Monteiro-Riviere [2006]. Penetration of intact skin by quantum dots with diverse physicochemical properties. Toxicol Sci 91(1):159–165.

Sager T, Porter D, Castranova V [2008]. Pulmonary response to intratracheal instillation of fine or ultrafine carbon black or titanium dioxide: Role of surface area. Toxicologist 102:A1491.

Sayes CM, Fortner JD, Guo W, Lyon D, Boyd AM, Ausman KD, Tao YJ, Sitharaman B, Wilson LJ, Hughes JB, West JL, Colvin VL [2004]. The differential cytotoxicity of water-soluble fullerenes. Nano Letters 4(10):1881–1887.

Sayes CM, Liang F, Hudson JL, Mendez J, Guo W, Beach JM, Moore VC, Doyle CD, West JL, Billups WE, Ausman KD, Colvin VL [2005]. Functionalization density dependence of single-walled carbon nanotubes cytotoxicity in vitro. Toxicol Lett 161(2):135–142.

Sayes CM, Wahi R, Kurian PA, Liu Y, West JL, Ausman KD, Warheit DB, Colvin VL [2006]. Correlating nanoscale titania structure with toxicity: a cytotoxicity and inflammatory response study with human dermal fibroblasts and human lung epithelial cells. Toxicol Sci 92(1):174–185.

Schneider T, Cherrie JW, Vermeulen R, Kromhout H [1999]. Conceptual model for assessment of dermal exposure. J Occup Environ Med 56:765–773.

Schneider T, Cherrie JW, Vermeulen R, Kromhout H [2000]. Dermal exposure assessment. Ann Occup Hyg 44(7):493–499.

Schulte P, Geraci C, Zumwalde R, Hoover M, Kuempel E [2008a]. Occupational risk management of engineered nanoparticles. J Occup Environ Hyg 5: 239–249.

Schulte PA, Trout D, Zumwalde R, Kuempel E, Geraci C, Castranova V, Mundt DJ, Mundt KA, Halperin WE [2008b]. Options for occupational health surveillance of workers potentially exposed to engineered nanoparticles: state of the science. J Occup Environ Med 50:517–526.

Seinfeld JA, Pandis SN [1998]. Atmospheric chemistry and physics. New York: John Wiley and Sons.

Semmler M, Seitz J, Erbe F, Mayer P, Heyder J, Oberdorster G, Kreyling WG [2004]. Long-term

Approaches to Safe Nanotechnology

clearance kinetics of inhaled ultrafine insoluble iridium particles from the rat lung, including transient translocation into secondary organs. Inhal Toxicol *16*(6–7):453–459.

Shalev I, Barker RL, McCord MG, Tucker PA, Lisk BR [2000]. Protective textile particulate penetration screening. Performance of protective clothing: 7th Symposium, ASTM STP 1386, West Conshohocken, PA: American Society for Testing and Materials, ASTM pp. 155–161.

Shiohara A, Hoshino A, Hanaki K, Suzuki K, Yamamoto K [2004]. On the cytotoxicity of quantum dots. Microbiol Immunol *48*(9):669–675.

Shvedova AA, Kisin ER, AR Murray, Gandelsman VZ, Maynard AD, Baron PA, Castranova V [2003]. Exposure to carbon nanotube material: assessment of the biological effects of nanotube materials using human keratinocyte cells. J Toxicol Environ Health *66*(20):1909–1926.

Shvedova AA, Kisin ER, Murray AR, Schwegler-Berry D, Gandelsman VZ, Baron P, Maynard A, Gunther MR, Castranova V [2004]. Exposure of human bronchial epithelial cells to carbon nanotubes cause oxidative stress and cytotoxicity. In: Proceedings of the Society for Free Radical Research Meeting, Paris, France: Society for Free Radical Research International, European Section, June 26–29, 2003.

Shvedova AA, Kisin ER, Mercer R, Murray AR, Johnson VJ, Potapovich AI, Tyurina YY, Gorelik O, Arepalli S, Schwegler-Berry D [2005]. Unusual inflammatory and fibrogenic pulmonary responses to single walled carbon nanotubes in mice. Am J Physiol Lung Cell Mol Physiol *289*(5):L698–708. Epub 2005 Jun 10.

Shvedova AA, Sager T, Murray A, Kisin E, Porter DW, Leonard SS, Schwegler-Berry D, Robinson VA, Castranova V [2007]. Critical issues in the evaluation of possible effects resulting from airborne nanoparticles. In: Monteiro-Riviere N and Tran L (eds). Nanotechnology: characterization, dosing and health effects. Philadelphia, PA: Informa Healthcare, pp. 221–232.

Shvedova AA, Kisin E, Murray AR, Johnson V, Gorelik O, Arepalli S, Hubbs AF, Mercer RR, Stone S, Frazer D, Chen T, Deye G, Maybnard A, Baron P, Mason R, Kadiiska M, Stadler K, Mouithys-Mickalad A, Castranova V, Kaagan VE [2008]. Inhalation of carbon nanotubes induces oxidative stress and cytokine response causing respiratory impairment and pulmonary fibrosis in mice. Toxicologist *102*:A1497.

Sriram K, Porter DW, Tsuruoka S, Endo M, Jefferson AM, Wolfarth MG, Rogers GM, Castranova V, Luster MI [2007]. Neuroinflammatory response following exposure to engineered nanomaterials. Toxicologist *96*:A1390.

Steenland K, Deddens J, Stayner L [1998]. Diesel exhaust and lung cancer in the trucking industry: exposure-response analyses and risk assessment. Am J Ind Med *34*(3):220–228.

Stefaniak AB, Hoover MD, Dickerson RM, Peterson EJ, Day GA, Breysse PN, Kent MS, Scripsick RC [2003]. Surface area of respirable beryllium metal, oxide, and copper alloy aerosols and implications for assessment of exposure risk of chronic beryllium disease. Am Ind Hyg Assoc J *64*:297–305.

Stevens G, Moyer E [1989]. Worst case aerosol testing parameters: I. sodium chloride and dioctyl phthalate aerosol filter efficiency as a function of particle size and flow rate. Am Ind Hyg Assoc J *50*(5):257–264.

Takagi A, Hirose A, Nishimura T, Fukumori N, Ogata A, Ohashi N, Kitajima S, Kanno J [2008]. Induction of mesothelioma in p53+/- mouse by intraperitoneal application of multi-wall carbon nanotube. J Toxicol Sci *33*(1):105–16.

Takenaka S, D Karg, C Roth, H Schulz, A Ziesenis, U Heinzmann, P Chramel, Heyder J [2001]. Pulmonary and systemic distribution of inhaled ultrafine silver particles in rats. Environ Health Persp *109*(suppl. 4):547–551.

The Royal Society, The Royal Academy of Engineering [2004]. Nanoscience and nanotechnologies. London, UK: The Royal Society and The Royal Academy of Engineering [www.nanotec.org.uk/finalReport.htm].

Thomas K, Aguar P, Kawasaki H, Morris J, Nakanish J, Savage N [2006]. Research strategies for safety evaluation of nanomaterials, part VIII: international efforts to develop risk-based safety evaluations for nanomaterials. Toxic Sci *92*(1):23–32.

Timonen KL, G Hoek, J Heinrich, A Bernard, B Brunekreef, K de Hartog, K Hameri, A Ibald-Mulli, A Mirme, A Peters, P Tiittanen, WB Kreyling, J Pekkanen [2004]. Daily variation in fine and ultrafine particulate air pollution and urinary concentrations of lung Clara cell protein CC16. Occup Environ Med *61*(11):908–914.

Tinkle SS, Antonini JM, Rich BA, Robert JR, Salmen R, DePree K, Adkins EJ [2003]. Skin as a route of exposure and sensitization in chronic beryllium disease. Environ Health Perspect *111*(9):1202–1208.

Törnqvist H, Mills NL, Gonzalez M, Miller MR, Robinson SD, Megson IL, Macnee W, Donaldson K, Söderberg S, Newby DE, Sandström T, Blomberg A [2007]. Persistent endothelial dysfunction in humans after diesel exhaust inhalation. Am J Respir Crit Care Med *176*(4):395–400.

Tran CL, Cullen RT, Buchanan D, Jones AD, Miller BG, Searl A, Davis JMG, Donaldson K [1999]. Investigation and prediction of pulmonary responses to dust, part II. In: Investigations into the pulmonary effects of low toxicity dusts. Contract Research Report 216/1999 Suffolk, UK: Health and Safety Executive.

Tran C, Buchanan LD, Cullen RT, Searl A, Jones AD, Donaldson K [2000]. Inhalation of poorly soluble particles. II. Influence of particle surface area on inflammation and clearance. Inhal Toxicol *12*(12):1113–1126.

TSI [2005]. Mechanisms of filtration for high efficiency fibrous filters. Application Note ITI–041, TSI Incorporated [www.tsi.com/AppNotes/appnotes.aspx?Cid=24 &Cid2=195&Pid=33&lid=439&file=iti_041#mech].

US DOE [2007]. Approach to Nanomaterial ES&H, U.S. Department of Energy's Nanoscale Science Research Centers. Washington, DC: U.S. Department of Energy.

VanOsdell DW, Liu BYH, Rubow KL, Pui DYH [1990]. Experimental-study of submicrometer and ultrafine particle penetration and pressure-drop for high-efficiency filters. Aerosol Sci Technol *12*(4):911–925.

Vaughan NP, Milligan BD, Ogden TL [1989]. Filter weighing reproducibility and the gravimetric detection limit. Ann Occup Hyg *33*(3):331–337.

Wang ZM, Gao P [2007]. A study on magnetic passive aerosol sampler for measuring aerosol particle penetration through protective ensembles. Unpublished paper presented at the 26th Annual Conference of the American Association for Aerosol Research. Reno, NV, September 24–28.

Wang L, Castranova V, Rojanasakul Y, Lu Y, Scabilloni J, Mercer RR [2008]. Direct fibrogenic effects of dispersed single walled carbon nanotubes on human lung fibroblasts. Toxicologist *102*:A1499.

Warheit DB, Laurence BR, Reed KL, Roach DH, Reynolds GA, Webb TR [2004]. Comparative pulmonary toxicity assessment of single-wall carbon nanotubes in rats. Toxicol Sci *77*:117–125.

Warheit DB, Webb TR, Sayes CM, Colvin VL, Reed KL [2006]. Pulmonary instillation studies with nanoscale TiO₂ rods and dots in rats: toxicity is not dependent upon particle size and surface area. Toxicol Sci *91*(1):227–236.

Warheit DB, Webb TR, Reed KL, Frerichs S, Sayes CM [2007]. Pulmonary toxicity study in rats with three forms of ultrafine-TiO₂ particles: differential responses related to surface properties. Toxicology *230*:90–104.

Wiebel A, Bouchet R, Boulch F, Knauth P [2005]. The big problem of particle size: a comparison of methods for determination of particle size in nanocrystalline anatase powers. Chem Mater *17*:2378–2385.

Woo KS, Chen DR, Pui DYH, Wilson WE [2001]. Use of continuous measurements of integral aerosol parameters to estimate particle surface area. Aerosol Sci Technol *34*:57–65.

Wood JP [2001]. Containment in the pharmaceutical industry. New York: Marcel Dekker, Inc.

Sources of Additional Information

Aitken RJ, Creely KS, Tran CL [2004]. Nanoparticles: an occupational hygiene review. HSE Research Report 274. UK Health and Safety Executive [www.hse.gov. uk/research/rrhtm/rr274.htm].

Balazy A, Podgórski A, Grado L [2004]. Filtration of nanosized aerosol particles in fibrous filters: I. Experimental results. Warsaw Poland: Warsaw University of Technology, Department of Chemical and Process Engineering.

Baron PA, Willeke K [2001]. Aerosol measurement. Principles, techniques and applications. New York: Wiley-Interscience.

Blake T, Castranova V, Schwegler-Berry D, Baron P, Deye GJ, Li CH, Jones W [1998]. Effect of fiber length on glass microfiber cytotoxicity. J Toxicol Environ Health *54*(Part A)(4):243–259.

Borm PJA [2002]. Particle toxicology: from coal mining to nanotechnology. Inhal Toxicology 14:311–324.

Byrd DM, Cothern CR [2000]. Introduction to risk analysis. Rockville, MD: Government Institutes.

Castranova V [1998]. Particles and airways: basic biological mechanisms of pulmonary pathogenicity. Appl Occup Environ Hyg 13(8):613–616.

Conhaim RL, Eaton A, Staub NC, Heath TD [1988]. Equivalent pore estimate for the alveolar-airway barrier in isolated dog lung. J Appl Physiol 64(3):1134–1142.

Driscoll KE, Costa DL, Hatch G, Henderson R, Oberdörster G, Salem H, Schlesinger RB [2000]. Intratracheal instillation as an exposure technique for the evaluation of respiratory tract toxicity: uses and limitations. Toxicol Sci 55(1):24–35.

EC [2004]. Nanotechnologies: a preliminary risk analysis on the basis of a workshop organized in Brussels, March 1–2, 2004, by the Health and Consumer Protection Directorate General of the European Commission. Brussels, Germany: European Commission [http://www.certh.gr/dat/7CF3F1C6/file.pdf?633556363252496250].

EPA [1992]. Guidelines for exposure assessment. Washington, DC: U.S. Environmental Protection Agency.

Faux SP, Tran CL, Miller BG, Jones AD, Monteiller C, Donaldson K [2003]. In vitro determinants of particulate toxicity: the dose-metric for poorly soluble dusts. Research report 154. Suffolk, UK: Health and Safety Executive.

Ferin J, Oberdörster G [1992]. Translocation of particles from pulmonary alveoli into the interstitium. J Aerosol Med 5(3):179–187.

Gibbs JP, Crump KS, Houck DP, Warren PA, Mosley WS [1999]. Focused medical surveillance: a search for subclinical movement disorders in a cohort of U.S. workers exposed to low levels of manganese dust. Neurotoxicol 20(2–3):299–313.

Hardman R [2006]. A toxicologic review of quantum dots: toxicity depends on physicochemical and environmental factors. Environ Health Perspect 114(2):165–172.

Hayes AW, ed. [2001]. Principles and methods of toxicology. 4th ed. Philadelphia, PA: Taylor and Francis.

Hoet PHM, Bruske-Hohlfeld I, Salata OV [2004]. Nanoparticles-known and unknown health risks. J Nanobiotech 2(12):1–15.

Hofmann W, Sturm R, Winkler-Heil R, Pawlak E [2003]. Stochastic model of ultrafine particle deposition and clearance in the human respiratory tract. Radiat Prot Dosim 105(1–4):77–80.

Hood E [2004]. Nanotechnology: looking as we leap. Environ Health Perspect 112(13):A741–A749.

Huckzko A, Lange H [2001]. Carbon nanotubes: experimental evidence for a null risk of skin irritation and allergy. Fullerene Sci Tech 9(2):247–250.

Huckzko A, Lange H, Calko E, Grubeck-Jaworska H, Droszez P [2001]. Physiological testing of carbon nanotubes: are they asbestos-like? Fullerene Sci Tech 9(2):251–254.

Keller A, Fierz M, Siegmann K, Siegmann HC, Filippov A [2001]. Surface science with nanosized particles in a carrier gas. J Vac Sci Technol A 19(1):1–8.

Klaassen CD, Amdur MO, Doull J [1986]. Casarett and Doull's toxicology: the basic science of poisons. New York: Macmillan Publishing Company.

Last JM [1983]. A dictionary of epidemiology. New York: Oxford University Press.

Lee KW, Liu BYH [1981]. Experimental study of aerosol filtration by fibrous filters. Aerosol Sci Technol 1(1):35–46.

Lee KW, Liu BYH [1982]. Theoretical study of aerosol filtration by fibrous filters. Aerosol Sci Technol 1(2):147–162.

Li N, Sioutas C, Cho A, Schmitz D, Misra C, Sempf J, Wang MY, Oberley T, Froines J, Nel A [2003]. Ultrafine particulate pollutants induce oxidative stress and mitochondrial damage. Environ Health Perspect 111(4):455–460.

Maynard AD [2001]. Experimental determination of ultrafine TiO_2 de-agglomeration in surrogate pulmonary surfactant—preliminary results. Inhaled Particles IX, Cambridge, UK: The British Occupational Hygiene Society.

Maynard AD [2002]. Experimental determination of ultrafine TiO_2 de-agglomeration in surrogate pulmonary surfactant—preliminary results. Ann Occup Hyg 46(Suppl 1):197–202.

NCI [1979]. Bioassay of titanium dioxide for possible carcinogenicity. Technical Report Series No. 97. Bethesda, MD: National Cancer Institute.

Nemmar A, Vanbilloen H, Hoylaerts MF, Hoet PHM, Verbruggen A, Nemery BS [2001]. Passage of intratracheally instilled ultrafine particles from the lung into the systemic circulation in hamster. Am J Respir Crit Care Med 164(9):1665–1668.

NFPA [1994]. Standard on protective ensembles for first responders to CBRN terrorism incidents. Quincy, MA: National Fire Protection Association.

Nichols G, Byard S, Bloxham MJ, Botterill J, Dawson NJ, Dennis A, Diart V, North NC, Sherwood JD [2002]. A review of the terms agglomerate and aggregate with a recommendation for nomenclature used in powder and particle characterization. J Pharm Sci 91(10):2103–2109.

Nel A, Xia T, Madler L, Li N [2006]. Toxic potential of materials at the nanolevel. Science 311:622–627.

NIOSH [1987]. NIOSH guide to industrial respiratory protection. Cincinnati, OH: U.S. Department of Health and Human Services, Centers for Disease Control and Prevention, National Institute for Occupational Safety and Health, DHHS (NIOSH) Publication No. 87–116, Appendix B.

NRC [1994]. Science and judgement in risk assessment. Washington D.C.: National Academy Press.

Oberdörster E [2004]. Manufactured nanomaterials (Fullerenes, C60) induce oxidative stress in brain of juvenile largemouth bass. Environ Health Perspect 112(10):1058–1062.

Oberdörster G, Ferin J, Finkelstein J, Soderholm S [1992]. Thermal-degradation events as health-hazards—Particle Vs gas-phase effects, mechanistic studies with particles. Acta Astronaut 27:251–256.

OSHA [1998]. Occupational Safety and Health Administration, OSHA 3143 (revised): industrial hygiene. Washington, DC: U.S. Department of Labor, Occupational Safety and Health Administration.

Patty [2000]. Patty's industrial hygiene and toxicology. New York: John Wiley and Sons.

Pflucker F, Wendel V, Hohenberg H, Gärtner E, Will T, Pfeiffer S, Wept F, Gers-Barlag H [2001]. The human stratum corneum layer: an effective barrier against dermal uptake of different forms of topically applied micronised titanium dioxide. Skin Pharmacol Appl Skin Physiol 14(Suppl 1):92–97.

Pui DYH, Chen D [2002]. Experimental and modeling studies of nanometer aerosol filtration. Sponsored by the Department of Energy. Minneapolis, MN: University of Minnesota, Particle Technology Laboratory. Grant No. DOEIDE–FG02–98ER1490.

Rao GVS, Tinkle S, Weissman DN, Antonini JM, Kashon ML, Salmen R, Battelli LA, Willard PA, Hoover MD, Hubbs AF [2003]. Efficacy of a technique for exposing the mouse lung to particles aspirated from the pharynx. J Toxicol Environ Health 66(15):1441–1445.

Renwick LC, Donaldson K, Clouter A [2001]. Impairment of alveolar macrophage phagocytosis by ultrafine particles. Toxicol Appl Pharmacol 172(2):119–127.

Roels H, Lauwerys R, Buchet JP, Genet P, Sarhan MJ, Hanotiau I, Defays M, Bernard A, Stanescu D [1987]. Epidemiologic survey among workers exposed to manganese—effects on lung, central-nervous-system, and some biological indexes. Am J Ind Med 11(3):307–327.

Schultz J, Hohenberg H, Plücker F, Gärtner E, Will T, Pfeiffer S, Wepf R, Wendel V, Gers-Barlag H, Wittern KP [2002]. Distribution of sunscreens on skin. Adv Drug Deliv Rev 54(Suppl 1):S157–S163.

Scott R [1997]. Basic concepts of industrial hygiene. Boca Raton, LA: Lewis Publishers.

Stefaniak AB, Hoover MD, Dickerson RM, Peterson EJ, Day GA, Breysse PN, Kent MS, Scripsick RC [2003]. Surface area of respirable beryllium metal, oxide, and copper alloy aerosols and implications for assessment of exposure risk of chronic beryllium disease. Am Ind Hyg Assoc J 64(3):297–305.

Thomas K, Sayre P [2005]. Research strategies for safety evaluation of nanomaterials, Part 1: evaluation the human health implications of exposure to nanoscale materials. Toxicol Sci 87(2):316–321.

Wang HC [1996]. Comparison of thermal rebound theory with penetration measurements of nanometer particles through wire screens. Aerosol Sci Technol 24(3):129–134.

Wang HC, Kasper G [1991]. Filtration efficiency of nanometer-size aerosol-particles. J Aerosol Sci 22(1):31–41.

Wennberg A, Iregren A, Struwe G, Cizinsky G, Hagman M, Johansson L [1991]. Manganese exposure in steel smelters a health-hazard to the nervous-system. Scand J Work Environ Health 17(4):255–262.

Appendix

Nanoparticle Emission Assessment Technique for Identification of Sources and Releases of Engineered Nanomaterials

1.0 Introduction

This appendix describes a technique that can be used by industrial hygienists for conducting initial workplace assessments for possible nanoparticle emissions. It allows a semi-quantitative evaluation of processes and tasks in the workplace where releases of engineered nanoparticles may occur. NIOSH uses several sampling approaches simultaneously with the goal of obtaining key physicochemical particle metrics: number concentration, qualitative size, shape, degree of agglomeration, and mass concentration of elemental constituents of interest.

2.0 Scope

Employers, workers, and researchers engaged in the production and use of engineered nanomaterials have expressed an interest in determining whether these nanomaterials are hazardous and if the potential for worker exposure exists. NIOSH has an active toxicology program to assess the potential hazards of engineered nanoparticles. Unfortunately these studies require long time periods and fall behind the pace of production and use of these nanomaterials. To assist in answering the latter of these questions, NIOSH established a nanotechnology field research team tasked with visiting facilities and collecting information about the potential for release of nanomaterials and worker exposure at those facilities. The initial challenges that the field research team encountered were: 1) determining which exposure metric (e.g., mass, particle number concentration, particle surface area) for engineered nanoparticles would provide a consistent body of knowledge to align with the toxicological results observed in experimental animal studies; and 2) selecting a sampling method based on metrics that were practical and would provide reproducible results. Engineered nanomaterials can be measured in the workplace using a variety of instrumentation including: condensation particle counter (CPC); optical particle counter (OPC); scanning mobility particle sizer (SMPS); electric low pressure impactor (ELPI); aerosol diffusion charger; and tapered element oscillating microbalance (TOEM), which vary in complexity and field portability. Unfortunately, relatively few of the above instruments are readily applicable to routine exposure monitoring due to non-specificity, lack of portability, difficulty of use, and high cost. NIOSH researchers have developed and used a field assessment strategy for determining exposures to engineered nanoparticles that could be adopted by other health and safety professionals in the evaluation of occupational exposures [Methner, et. al. 2007; Methner, 2008].

Since there are currently no exposure limits specific to engineered nanomaterials, this technique is used to determine whether airborne releases of engineered nanomaterials occur. This assessment, which compares particle number concentrations and relative particle size at the potential emission source to background particle number concentrations and particle size, provides a semi-quantitative means for determining the effectiveness of existing control measures in reducing engineered nanoparticle exposures. This procedure utilizes portable direct-reading instrumentation supplemented by filter-based air samples (source-specific and personal breathing zone [PBZ]). The use of filter samples is crucial for particle identification because direct-reading instruments used for determining particle number concentrations are incapable of identifying the composition of the particles.

3.0 Summary of the On-Site Initial Assessment

The initial assessment uses a combination of direct-reading, handheld instruments (CPC and OPC) and filter-based sampling (e.g. 37-mm diameter filter cassettes) for subsequent chemical and microscopic analyses (Figure 1). This semi-quantitative approach was first described by Maynard et al. [2004] and NIOSH has adopted a similar approach. The technique includes determining particle number concentration using direct-reading, handheld particle counters at potential emission sources and comparing those data to background particle number concentrations. If elevated concentrations of suspected nanoparticles are detected at potential emission sources, relative to the background particle number concentrations, then a pair of filter-based, source-specific air samples are collected with one sample analyzed by transmission electron microscopy (TEM) or scanning electron microscopy (SEM) for

particle identification and characterization, and the other used for determining the elemental mass concentration (Figure 2). A second pair of filter-based air samples may also be collected in the personal breathing zone of workers. Breathing zone samples are analyzed in the same manner as the area air samples (i.e., by TEM and elemental mass).

4.0 Air Sampling Instrumentation and Filter Media Used in the Initial Assessment

The following instrumentation is used by NIOSH; however, use does not constitute endorsement.

4.1 TSI model 3007 (or model 8525) (TSI Inc, Shoreview, MN), handheld condensation particle counter (CPC), which uses isopropanol to condense on particles so they can be counted

The TSI units provide a non-specific measure of the total number of particles independent of chemical identity per cubic centimeter of air (P/cm^3). The measureable range is between 10–1,000 nm for model 3007, or between 20–1,000 nm for model 8525. The range of detection for these instruments is reported by the manufacturer to be 0–100,000 P/cm^3.

4.2 ART Instruments Hand Held Particle Counter (HHPC-6, ART Instruments, Grants Pass, Oregon), which operates on optical counting principles using laser light scattering.

The HHPC-6 optical particle counter (OPC) can measure the total number of particles per liter (P/L)

of air independent of chemical identity within six specific size ranges. The OPC used by the NIOSH field research team provides particle counts in the following size cut-points: 300 nm; 500 nm; 1,000 nm; 3,000 nm; 5,000 nm; and 10,000 nm. The range of detection for this instrument is reported by the manufacturer to be 0–70,000 P/L. Different manufacturers' OPCs may have slightly different particle size ranges and could be substituted.

4.3   Appropriate air sampling filter media (e.g. mixed cellulose ester, quartz fiber filter) are selected depending on nanoparticle type and desired analytical information (e.g., determination of particle morphology using TEM or SEM, elemental analysis for metals, elemental analysis for carbon)

4.4   Air sampling pumps capable of sampling at high flow rates (e.g., 7 liters per minute or other flow rate depending upon the duration of the task and the appropriate NIOSH method, if a method is available)

4.5   Sampling pump flow calibrator

4.6   If desired, personal cascade impactor or respirable cyclone (see 5.3.3)

4.7   If desired, cassette conductive cowl (see 5.3.3)

4.8   Optional research-grade particle analyzers for expanded surveys (see 5.6.1)

4.9   Optional surface sampling supplies such as substrate (e.g., Ghost Wipes™), disposable 10 cm × 10 cm templates, sterile containers, and

nitrile gloves for handling media (see 5.6.2)

5.0 Evaluation of Potential Releases of Engineered Nanomaterials

5.1 Identify Potential Sources of Emissions

The overall purpose of this step is to develop a list of target areas and tasks that will be evaluated with the particle analyzers.

The initial assessment involves identifying the potential sources of engineered nanomaterial emissions by reviewing the type of process, process flow, material inputs and discharges, tasks, and work practices. When available, literature (e.g., MSDS, records of feedstock materials) is reviewed to gain an understanding of the engineered nanomaterials being produced or used, including their physicochemical properties such as size, shape, solubility, and reactivity. Once the potential sources of emissions have been identified from the process review, the industrial hygienist (or other qualified person):

- Conducts an observational walk-through survey of the production area and processes to locate potential sources of emissions.

- Determines the frequency and duration of each operation and the type of equipment used for handling and containment of the material.

- Determines the presence/absence of general and local exhaust ventilation and other engineering controls. (This initial assessment includes identifying points of potential system failure that could result in emission from the containment/control system [e.g., hole in duct, deteriorated sealing gasket]).

- Determines the process points where containment is deliberately breached (e.g., opening system for product retrieval or for cleaning).

5.2 Conduct Particle Concentration Sampling

5.2.1 Background measurements

Determining the contribution of background particle concentrations on measurements made for the particles of interest (e.g., engineered nanoparticles) is an important evaluation of assessing the possible airborne release of engineered nanoparticles.

Ideally, during the initial assessment, the industrial hygienist (or other qualified person), will determine the average airborne particle concentration at various processes and adjacent work areas with the CPC and OPC *before* the processing or handling of nanomaterials begins. If the background particle concentrations are high (values are relative and will vary with processes and facilities), an assessment is made as to whether there may be a source of incidental nanoparticles in the area. Incidental nanoparticles may be generated from a variety of sources,

including vacuum pumps, natural gas heating units, gasoline/propane/diesel powered fork lift trucks, or other combustion activities such as welding, soldering, or heat-sealing. The CPC and OPC can be used to check these sources for incidental nanoparticle releases. Outdoor or re-circulated air supply from the building ventilation system should also be considered as a possible source of nanoparticles [Peters et al. 2006].

Measurements of background particle concentrations are repeated after the active processing, manufacturing, or handling of the nanomaterial has ended. An average background concentration is then computed and subtracted from the measurements made during processing, manufacturing, or the handling of engineered nanomaterials. This approach is acceptable only if background particle counts remain relatively stable throughout the measurement period and particle emissions from the process under investigation are sufficiently elevated above background. For other situations, correcting for particle background concentrations becomes more complex requiring additional sampling over an extended time period to determine the source(s) and magnitude of background particle concentrations. This type of evaluation is generally outside the scope of the initial assessment described here.

5.2.2 Area sampling

Once initial background particle concentrations have been determined, measurements of airborne particle concentrations and size ranges are made with the CPC and OPC simultaneously at locations near the suspected or likely emission source (e.g., opening a reactor, handling product, potential leak points in the ventilation system). Airborne particle concentrations are determined before, during, and after each task or operation to identify those factors (e.g., controls, worker interaction, work practices) that may affect airborne particle concentrations. This information is used to identify processes, locations, and personnel for filter-based air sampling (5.3).

5.3 Conduct Filter-based Area and Personal Air Sampling

5.3.1 *Area air sampling*

A pair of filter-based, air samples are collected at process/task locations and/or workers engaged in process operations where suspected engineered nanomaterial emissions may occur, based on air sampling results using the CPC and OPC.

Filter-based area air samples provide more specific information on the engineered nanomaterial of interest (e.g., size, shape, mass). The pair of air samples includes one sample analyzed for elemental mass and one sample analyzed by electron microscopy. For example, one sample might be collected for metals determination (e.g., NIOSH Method 7300, 7303) or elemental carbon (e.g., NIOSH Method 5040) depending on the composition of the engineered nanomaterial. The other sample would be collected for particle characterization

(e.g., size, shape, dimension, degree of agglomeration) by TEM or SEM using the measurement techniques specified in NIOSH Methods 7402, 7404, or other equivalent methods [NIOSH 1994].

The source-specific air samples are collected as close as possible to the suspected emission source but outside of any existing containment, to increase the probability of detecting any possible release of engineered nanomaterials. Sampling duration generally matches the length of time in which the potential exposure to the engineered nanomaterial exists at the task or specific process. In cases where the duration of the tasks associated with the potential airborne release of nanomaterials is short (e.g., minutes), a relatively high air sampling flow rate may be required (approximately 7 liters per minute) to ensure adequate particle loading on the filter media. If specific information is desired on the worker's potential exposure to the engineered nanomaterial then PBZ samples should be collected using the two- sample filter-based sampling strategy described above.

If the particle number concentrations (using CPC or OPC) are substantially high, then shorter sampling times for the TEM or SEM sample may be necessary to avoid overloading the filter and interfering with particle characterization. The specific sampling time should be based on direct-reading instrument results and professional judgment of the industrial hygienist. In general, filter samples are collected for the duration of a given task, normally 15–30 minutes. If the direct-reading instruments indicate a high particle number concentration the sampling time can be shortened to 5–10 minutes, or both a short- and long-duration sample may be collected to ensure an adequate sample for electron microscopy analysis. See Table 1 for additional sampling time guidance. However, the sampling times in Table 1 were based on collection of asbestos fibers by NIOSH Method 7402 and may not be applicable for much smaller engineered nanoparticles. See Figures 3–5 for example TEM micrographs.

A minimum of 2 background filter samples are collected distant from the potential sources of engineered nanoparticle exposure to serve as an indicator of ambient particle identification and concentration.

5.3.2 *Personal air samples*

When possible, personal breathing zone (PBZ) air samples are collected on workers likely to be exposed to engineered nanomaterials (e.g., engaged in active handling of nanomaterials or operating equipment previously identified as emitting nanoparticles). If measurements obtained with the CPC and OPC indicate that nanoparticles are being emitted at a specific process where a worker is located, then the collection of PBZ samples may be warranted.

PBZ samples are analyzed in the same manner as the area air samples (i.e., by TEM and elemental mass). It may be necessary to collect samples at a relatively high flow rate (e.g., 7 liters per minute) if the duration of the task and the resulting potential exposure is short.

5.3.3 *Optional sample collection*

In the event that measurements made by the OPC indicate a large fraction (over 50%) of particles exceeding 1,000 nm in size, the use of a personal cascade impactor or respirable cyclone sampler in tandem with a filter-based air sampling cassette may be required for both the mass and TEM/SEM analyses to eliminate large particles that may interfere with analysis and be of limited interest. The use of an impactor or cyclone will require using a flow rate appropriate for the particle cut size and is usually in the range of 1.7–2.5 liters per minute. Open-face, and impactor or cyclone samples may be collected side by side to allow a more thorough interpretation of analytical results. Additionally, if it is anticipated that the nanoparticles of interest will have a tendency to be electrostatically attracted to the sides of the plastic air sampling cassette, a conductive cowl may be necessary to eliminate particle loss and subsequent underestimation of the airborne nanoparticle concentration. The use of a personal cascade impactor, respirable cyclone, or conductive cowl is made at the discretion of the industrial hygienist (or other qualified person).

If the facility is manufacturing or using TiO_2, then the sampling should include the sampling recommendations found in the NIOSH *Draft Document: Evaluation of Health Hazard and Recommendations for Occupational Exposure to Titanium Dioxide* (www.cdc.gov/

niosh/review/public/TiO2/default.html),which recommends collecting a mass-based airborne measurement using NIOSH Method 0600.

5.4 Quality Assurance and Quality Control

To ensure valid emission measurements, the following quality assurance and control steps should be taken:

- Use factory calibrated direct-reading particle analyzers

- Perform daily zero-checks on all particle counters before each use

- Calibrate pumps before and after each sampling day

- Submit for analysis any process, background, and bulk material samples along with field and media blanks to a laboratory accredited by the American Industrial Hygiene Association (AIHA)

5.5 Data Interpretation

Since the size of airborne engineered nanoparticles and the degree of agglomeration may be unknown at the time of sample collection, the use of direct-reading, particle sizing/counting instruments may provide a semi-quantitative indication of the magnitude of potential emissions, provided background particle number subtraction can be successfully accomplished. The particle number concentration measurements taken with the CPC and OPC will provide a measurement of particles larger than the ASTM definition of nanoparticles (1–100 nm) [ASTM 2006]. However, the two particle counters can be used simultaneously to obtain a semi-quantitative size differential evaluation of the aerosol being sampled. The CPC provides a measure of total particles per cm^3 in the size range of 10–1,000 nm (or 20–1,000 nm). The OPC provides the total number of particles per liter of air within six specific size ranges: 300 nm; 500 nm; 1,000 nm, 3,000 nm, 5,000 nm and > 10,000 nm. If necessary, the data from the CPC and OPC can be used together to determine the number concentration of nanoscale particles. For example, a high particle number concentration on the CPC, in combination with a high particle number concentration in the small size ranges (300–500 nm) on the OPC, may indicate the possible presence of nanoscale particles. Conversely, a low CPC particle number concentration, in combination with a high OPC particle number concentration in the larger size ranges (> 1,000 nm) may indicate the presence of larger particles and/or engineered nanoparticle agglomerates. These assumptions of nanoparticles versus larger particles and/or nanoparticle agglomerates may be verified by TEM or SEM analysis.

5.5.1 Selectivity

Selectivity is a critical issue when characterizing exposure using airborne particle number concentration. Airborne nanoparticles are present in many workplaces and often originate from multiple sources such as combustion, vehicle emissions, and infiltration of outside air. Particle counters are generally not selective to particle source or composition, making it difficult to differentiate between incidental and process-related nanoparticles using number concentration alone. The CPC and OPC are used to identify sources of nanoparticles and the filter-based samples are used to verify the size, shape, and chemical composition of the nanoparticles with the goal of differentiating between incidental and engineered nanoparticles.

5.5.2 *Limitations*

The exposure assessment technique does have some limitations including:

- Although this issue is not unique to particle number concentration measurements, orders of magnitude difference can exist in aerosol number concentrations, depending on the number and types of sources of particle emissions. Monitoring over several days and during different seasons can provide a better understanding of the variability that might exist in airborne particle number concentrations found in background measurements and in measurements made at sources where engineered nanomaterials are handled.

- The upper dynamic range of the CPC is 100,000 P/cm^3. A dilutor, consisting of a modified HEPA filter cartridge placed upstream of the inlet, can extend the range of the CPC when

particle number concentrations are greater than 100,000 P/cm^3 [Peters et al. 2006; Heitbrink et al. 2007; Evans et al. 2008].

- The analysis of air samples by TEM or SEM with energy dispersive X-ray spectrometry can provide information on the elemental composition of the nanomaterials. However, TEM and SEM analysis can be compromised if there is particle overload on the filter. Alternatively, if the loading is too sparse, an accurate assessment of particle characteristics may not be possible (see 5.3.1).

- Note that area samples are collected as closely as possible to the source of emission to allow for more accurate determination of a nanoparticle release and to identify locations most likely to result in worker exposure. **Therefore, results from this type of sampling should not be interpreted as representative of worker exposure.** However, samples collected in such a fashion should serve as an indicator of material release and the possible need for controls.

5.6 Expanded Research (In Depth Assessments)

5.6.1 *Research instrumentation*

A major obstacle in conducting more specific measurement of engineered nanomaterials in the workplace is a lack of field-portable instruments that can be easily maneuvered within a facility or easily worn by a worker to provide an indication of PBZ exposure. Additionally, there is no single instrument capable of measuring the numerous potential exposure metrics associated with engineered nanomaterials (e.g., number concentration, surface area, size, shape, mass concentration) [Maynard and Aitken 2007]. Although the following instruments lack field portability and ease of use, they can measure many of the desirable exposure metrics and provide information about the particle size distribution. These research-grade particle analyzers are not usually part of the initial assessment but are used when additional knowledge about the nanoscale particle temporal or spatial exposure variation or size distribution is desired.

5.6.1.1 *Particle Surface-Area Analyzers*

Toxicology studies have indicated that surface area of nanoparticles may be an important exposure dose metric. Portable aerosol diffusion chargers may be used to provide estimates of external aerosol surface area when airborne particles are smaller than 100 nm in diameter, but these may tend to overestimate external surface area when particles are larger than 100 nm in diameter. These instruments are based on diffusion charging followed by detection of the charged aerosol using an electrometer.

The TSI Aerotrak™ 9000 Nanoparticle Aerosol Monitor does not measure total active surface area but indicates the surface area of particles which may be deposited in the lung in units of square micrometers per cubic centimeter,

corresponding to either the tracheobronchial or alveolar regions of the lung. The Ecochem DC 2000-CE measures the total particle surface area. These devices are currently being evaluated as part of the process used by NIOSH to conduct initial assessments. These particle surface analyzers are used as area samplers.

5.6.1.2. *Scanning Mobility Particle Sizer*

More specific depictions of particles by size (diameter) and number can greatly improve the ability to evaluate possible releases of engineered nanoparticles. One particular instrument, the Scanning Mobility Particle Sizer (SMPS) measures particle diameters from 2.5–1,000 nm and can display data as a size and number distribution using up to 167 size channels. The SMPS is widely used as a research tool for characterizing nanoscale aerosols. The SMPS employs a continuous, fast-scanning technique to provide high-resolution measurements. However, the SMPS may take 2–3 minutes to scan which may not be useful for the process screening in workplaces with highly variable aerosol size distributions. Its applicability for use in the workplace may be limited because of its size, cost, and use of an internal radioactive source.

The Fast Mobility Particle Sizer (FMPS) is similar to the SMPS but has a much faster response time (approximately 1 second). However, because it has fewer particle size channels, it does not include the same level of detail on particle size distributions that can be determined with the SMPS.

The FMPS and SMPS are used as area samplers.

5.6.1.3 *Low Pressure Impactors*

The Electrical Low Pressure Impactor (ELPI) combines diffusion charging and a cascade impactor to provide aerosol size distributions by aerodynamic diameter as determined real time by mass and number collected on a series of plates.

Low pressure cascade impactors offer the ability to size particles and then conduct secondary analyses (e.g., metals analysis). However, these instruments are sensitive to harsh field conditions and are not considered portable. The ELPI is used as an area sampler.

5.6.1.4 *Tapered Element Oscillating Microbalance*

The tapered element oscillating microbalance (TEOM) is commonly used for sampling aerosols less than 1 μm in diameter, however, the sampling inlet can be set to select different size fractions. The TEOM determines mass by detecting a change in vibration frequency across a particle-collecting substrate. The TEOM can be configured to provide size-differentiated mass measurements and is used as an area sampler.

5.6.2 *Surface sampling*

Surface sampling to detect the presence of engineered nanomaterials is not routinely part of the initial assessment but may be conducted to determine if surface contamination exists. Surface sampling does not provide size-specific information but may be useful for determining whether engineered nanomaterials have migrated away from active production or handling areas and have contaminated nonproduction work areas. The decision to collect surface samples is made in the field at the discretion of the industrial hygienist (or other qualified person), and is dependent on direct observation and the nanomaterial of interest. For example, surface sampling was completed at a quantum dot facility after observing dusty surfaces in areas adjacent to the production area. In order to determine if the dust was contaminated with quantum dots, surface samples were collected and analyzed for the chemical components of the quantum dots produced by that facility.

Surface wipe samples are collected using a pre-moistened substrate such as Ghost Wipe™ towelettes in accordance with NIOSH Method 9102 for elements or the NIOSH method for specific elements (e.g., NIOSH Method 9100 for lead). When collecting wipe samples, the following steps should be followed:

- Don a pair of nitrile disposable gloves

- Wipe the surface within a disposable 10 cc × 10 cc template using four horizontal s-shaped strokes

- Fold the exposed side of the wipe in and wiping the same area with four vertical s-shaped strokes

- Fold the wipe, exposed side in, and placing it into a sterile container

Gloves and template are discarded after each sample collection to eliminate the possibility of cross-contaminating successive samples. Wipe samples may be collected from undisturbed horizontal surfaces throughout the facility at locations suspected to be contaminated and in areas expected to be free of engineered nanomaterials. Wipe samples are analyzed following the appropriate NIOSH method for the chemical substance of interest.

6.0 Conclusions

The NIOSH initial assessment technique uses complimentary approaches to semi-quantitatively evaluate the potential releases of engineered nanoparticles. Two different particle counters are used in a parallel and differential manner to evaluate the total particle number relative to background and the relative size distribution of the particles. If this initial evaluation indicates an elevated number of small particles, which could potentially be the engineered nanoparticle of interest, then the particle counters are used to detect the source of the emissions. If nanoparticles are found and determined to be emitted from a specific process (versus background incidental nanoscale particles), then additional samples are collected for qualitative measurement of particle size and shape, (by TEM or SEM analysis) and for determination of elemental mass concentration (by chemical analysis).

The initial assessment technique is useful for determining whether airborne releases of engineered nanomaterials are occurring at potential emission sources. This assessment provides a semi-quantitative means for determining whether existing measures are adequate for controlling nanomaterial emissions or if additional controls may be required.

The NIOSH emission assessment technique may be useful to health and safety professionals who are interested in determining whether release of nanomaterials occurs in the workplace. Where possible, use of the technique should be repeated in workplaces of interest to gain a better understanding of the daily fluctuations in airborne exposures at processes and tasks in which engineered nanomaterials occur and for determining potential sources of background particle number concentrations. A more systematic and routine assessment of the workplace can provide more definitive information on the performance of control measures and if additional actions are needed to reduce worker exposure.

The initial assessment technique can be expanded or modified to determine additional metrics (Figure 6). Research initiatives addressing more comprehensive process monitoring, particle metrics, personal exposure monitoring, and method/approach development and validation are currently underway within NIOSH. As this information becomes available, revisions to the Approaches to Safe Nanotechnology document will be made.

Information about contacting the nanotechnology field research team is available at: [www.cdc.gov/niosh/

Appendix

docs/2008-121], see the Fact Sheet: NIOSH Nanotechnology Field Research Effort [NIOSH 2008].

7.0 References

ASTM International [2006] ASTM-E2456–06 Standard terminology relating to nanotechnology

Evans DE, Heitbrink WA, Slavin TJ, Peters TM [2008]. Ultrafine and respirable particles in an automotive grey iron foundry. Ann Occup Hyg 52(1):9–21.

Heitbrink WA, Evans DE, Peters TM, Slavin TJ [2007]. The characterization and mapping of very fine particles in an engine machining and assembly facility. J Occup Environ Hyg 4:341–351.

NIOSH [2008]. NIOSH Fact Sheet: NIOSH Nanotechnology Field Research Effort, Cincinnati, OH: U.S. Department of Health and Human Services, Centers for Disease Control and Prevention, National Institute for Occupational Safety and Health. DHHS (NIOSH) Publication No. 2008–121 [http://www.cdc.gov/niosh/docs/2008-121/].

NIOSH [1994]. NIOSH manual of analytical methods (NMAM®), 4th ed. By Schlecht PC, O'Conner PF, eds. Cincinnati, OH: U.S. Department of Health and Human Services, Centers for Disease Control and Prevention, National Institute for Occupational Safety and Health. DHHS (NIOSH) Publication No. 94–113 [www.cdc.gov/niosh/nmam/].

Maynard A, Aitken R [2007]. Assessing exposure to airborne nanomaterials: current abilities and future requirements. Nanotoxicology 1(1):26–41.

Maynard A, Baron P, Foley M, Shvedova A, Kisin E, Castranova V [2004] Exposure to carbon nanotube material: Aerosol release during the handling of unrefined single walled carbon nanotubes material. J Toxicology. Environmental Health Part A, 67: 1; 87 107.

Methner MM, Birch ME, Evans DE, Ku BK, Crouch KG, Hoover MD [2007]. Mazzukelli LF, ed: Case study: Identification and characterization of potential sources of worker exposure to carbon nanofibers during polymer composite laboratory operations. J Occup Environ Hyg 4(12): D125–D130.

Methner M [2008]. Engineering case reports. Old L, ed. Effectiveness of local exhaust ventilation in controlling engineered nanomaterial emissions during reactor cleanout operations. J Occup Environ Hyg 5(6): D63–D69.

Peters T, Heitbrink W, Evans D, Slavin T, Maynard A [2006]. The mapping of fine and ultrafine particle concentrations in an engine machining and assembly facility. Ann Occup Hyg 50(3):1–9.

Table 1. Approximate sampling times for TEM grid based on particle number concentrations[*].

| | TEM grid | Open-faced cassettes | | |
		25-mm	37-mm	47-mm
Diameter (mm)	3.0	25.0	37.0	47.0
Effective diameter (mm)	3.0	22.2	34.2	44.2
Effective collection area (mm^2)	7	385	916	1531
Flow (L/min)	0.1	7	7	7
Desired Loading (#/mm^2)	1.E+06	1.E+06	1.E+06	1.E+06

Air concentration (#/cm^3)	Time (min)			
250	282.7	220.2	523.4	874.8
500	141.4	110.1	261.7	437.4
1,000	70.7	55.0	130.8	218.7
2,000	35.3	27.5	65.4	109.4
4,000	17.7	13.8	32.7	54.7
8,000	8.8	6.9	16.4	27.3
16,000	4.4	3.4	8.2	13.7
32,000	2.2	1.7	4.1	6.8
64,000	1.1	0.9	2.0	3.4
128,000	0.6	0.4	1.0	1.7

[*]NIOSH NMAM Method 7402 Asbestos by TEM and personal communication with Dr. Aleksandr Stefaniak (NIOSH)

Figure 1. A demonstration of the initial assessment technique with side-by-side sampling using (from left to right) the OPC, co-located open-face filter cassettes, and the CPC: examples of PBZ and source-specific filter-based sampling setup.

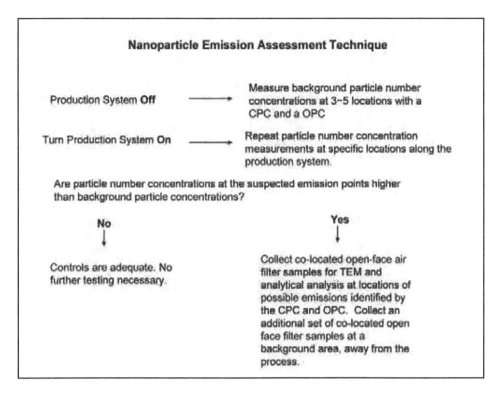

Figure 2. Summary of the initial assessment technique

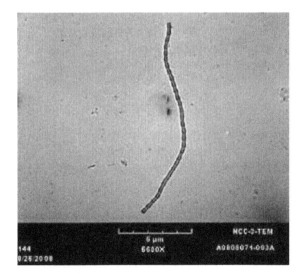

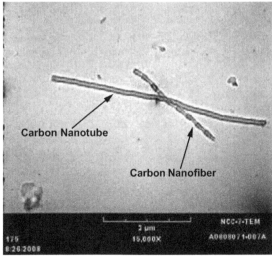

Figure 3. Electron microscopy micrograph of a carbon nanofiber

Figure 4. Electron microscopy micrograph of a carbon nanofiber and carbon nanotube

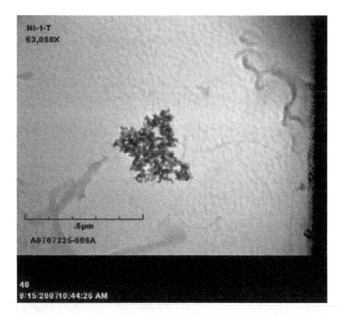

Figure 5. Electron microscopy micrograph of an agglomerated nanoparticle of nickel oxide

Expanded Nanoparticle Assessments

- Does the OPC indicate the majority of particles in the 1.0 um or larger size?
  - Yes. Use a personal cascade impactor or respirable cyclone in front of the air filters collected for TEM/SEM and mass. Collect a second set without the impactor/cyclone for comparison.

- Is there visible evidence of dust from the process on undisturbed horizontal surfaces?
  - Yes. Collect surface samples as per standard method (e.g. NIOSH Method 9102 or 9100). Select the appropriate analytical marker.

- Do you need to know the particle size distribution for nanoparticles <300 nm, or any additional temporal or spatial information?
  - Yes. Use advanced instrumentation including the SMPS, ELPI, BET, or TOEM.

Figure 6. Considerations for expanded nanomaterial assessments

Made in the USA
Las Vegas, NV
24 February 2021